Blockchain Technology

Internet of Everything (IoE)

Series Editors:
Vijender Kumar Solanki, Raghvendra Kumar, and Le Hoang Son

IoT
Security and Privacy Paradigm
Edited by Souvik Pal, Vicente Garcia Diaz, and Dac-Nhuong Le

Smart Innovation of Web of Things
Edited by Vijender Kumar Solanki, Raghvendra Kumar, and Le Hoang Son

Big Data, IoT, and Machine Learning
Tools and Applications
Rashmi Agrawal, Marcin Paprzycki, and Neha Gupta

Internet of Everything and Big Data
Major Challenges in Smart Cities
*Edited by Salah-ddine Krit, Mohamed Elhoseny, Valentina Emilia Balas,
Rachid Benlamri, and Marius M. Balas*

Bitcoin and Blockchain
History and Current Applications
*Edited by Sandeep Kumar Panda, Ahmed A. Elngar, Valentina
Emilia Balas, and Mohammed Kayed*

Privacy Vulnerabilities and Data Security Challenges in the IoT
Edited by Shivani Agarwal, Sandhya Makkar, and Tran Duc Tan

Handbook of IoT and Blockchain
Methods, Solutions, and Recent Advancements
*Edited by Brojo Kishore Mishra, Sanjay Kumar Kuanar,
Sheng-Lung Peng, and Daniel D. Dasig, Jr.*

Blockchain Technology
Fundamentals, Applications, and Case Studies
Edited by E. Golden Julie, J. Jesu Vedha Nayahi, and Noor Zaman Jhanjhi

Data Security in Internet of Things Based RFID and
WSN Systems Applications
Edited by Rohit Sharma, Rajendra Prasad Mahapatra, and Korhan Cengiz

For more information about this series, please visit: https://www.crcpress.com/
Internet-of-Everything-IoE-Security-and-Privacy-Paradigm/book-series/
CRCIOESPP

Blockchain Technology
Fundamentals, Applications, and Case Studies

Edited by
E. Golden Julie, J. Jesu Vedha Nayahi, and
Noor Zaman Jhanjhi

CRC Press
Taylor & Francis Group
Boca Raton London New York

CRC Press is an imprint of the
Taylor & Francis Group, an **Informa** business

First edition published 2021
by CRC Press
6000 Broken Sound Parkway NW, Suite 300, Boca Raton, FL 33487-2742

and by CRC Press
2 Park Square, Milton Park, Abingdon, Oxon, OX14 4RN

Library of Congress Cataloging-in-Publication Data
Names: Julie, Golden, 1984- editor. | Nayahi, J. Jesu Vedha, editor. |
Zaman, Noor, 1972- editor.
Title: Blockchain technology : fundamentals, applications, and case studies /
edited by E. Golden Julie, J. Jesu Vedha Nayahi, and Noor Zaman.
Other titles: Blockchain technology (CRC Press)
Description: First edition. | Boca Raton, FL : CRC Press, 2021. |
Series: Internet of everything (IoE). Security and privacy paradigm | Includes
bibliographical references and index.
Identifiers: LCCN 2020024166 (print) | LCCN 2020024167 (ebook) |
ISBN 9780367431372 (hardback) | ISBN 9781003004998 (ebook)
Subjects: LCSH: Blockchains (Databases)
Classification: LCC QA76.9.B56 B56286 2021 (print) | LCC QA76.9.B56 (ebook) |
DDC 005.74—dc23
LC record available at https://lccn.loc.gov/2020024166
LC ebook record available at https://lccn.loc.gov/2020024167

ISBN: 978-0-367-43137-2 (hbk)
ISBN: 978-0-367-61708-0 (pbk)
ISBN: 978-1-003-00499-8 (ebk)

DOI: 10.1201/9781003004998

Typeset in Times
by codeMantra

Contents

Preface

The purpose of this edited book is to present the detailed exploration of adaption and implementation of blockchain technologies in the real-world business applications. Blockchain is getting momentum in all the sectors that transact in huge numbers every day. This book covers all the aspects of the blockchain with complete 360-degree view spectrum, and it can be used at basic and intermediary levels by computer science postgraduate students, researchers, and practitioners. It presents the rapid advances made in the existing business model by applying the blockchain techniques.

Chapter 1 covers survey on blockchain and Internet of Things. This chapter focuses on traditional design of a blockchain, smart energy, smart city, smart green house, data protection, transportation, and industry IoT with detailed architecture. This chapter concludes with discussion on benefits of blockchain.

Chapter 2 aims to review protocols and standards of blockchain. It deals with key elements of blockchain technologies, important features of blockchain protocols, consensus algorithms in blockchain, importance of standards, and finally blockchain standards framework.

Chapter 3 discusses about blockchain technology. It provides an insight into architecture, use cases, and its application with industrial IoT and Big Data. This chapter deals with architectures for blockchain implementation, followed by use cases for blockchain implementation, IoT, and artificial intelligence.

Chapter 4 covers IoT and blockchain. This chapter discusses about the working principle of IoT with its characteristics, applications, and their pros and cons. It also deals with the blockchain–IoT convergence and its prominent applications as well as challenges.

Chapter 5 deals with research issues and solution in blockchain healthcare applications. Various research challenges and its solutions in healthcare sector have been discussed. It focuses mainly on the necessity of designing a secured healthcare network model for medical data exchange between various participants. It explores the emerging applications in decentralized ledger technology that addresses the need of health data to be authenticated, distributed, and immutable.

Chapter 6 covers improving security on blockchain and its integration with IoT. New technologies like cryptographic blockchain algorithms are suggested by the researchers to improve security of miners and participants. This chapter focuses on cryptographic algorithms needed to secure BIOT, delegated and authenticated consensus algorithms for different types of blockchains, secure designing of cryptographic digital ledger and revocation ledger, public/private cryptographic algorithms, hashing algorithms, fault-tolerance systems, and new access control list.

Chapter 7 deals with blockchain-based integrated digital health record model for health information exchanges. It discusses the importance of blockchain-based health care system that integrates patients with personal data, electronic health data, hospitals, physicians, insurance companies, and drug industry to improve health record management, enhance insurance claim process, and accelerate clinical/biomedical research.

Chapter 8 discusses bitcoin features, bitcoin transactions, bitcoin mining, and the merits of using bitcoin. The features, transactions, and merits of other protocols such as Ethereum, Ripple, and Litecoin are also explained.

Chapter 9 focuses on FGUGChain framework, which is a novel method to incorporate privacy in blockchain with the help of secure computation. The fundamentals of blockchain, secure computation, and the research directions on blockchain are explained.

Chapter 10 deals with blockchain technology for the removal of information silo by building an information network. The concept of digital block information systems is explained and the experimental results show that the proposed work removes information silo as well as ensures nonrepudiation of information.

Chapter 11 explains the application of blockchain in various fields such as food supply chain management, digital India, securing Electronic Health Records, and maintaining the Know Your Customer (KYC) data.

Chapter 12 discusses the applicability of blockchain in cryptocurrency. It details features and type of cryptocurrency as well as processing of transactions using blockchain.

Chapter 13 explains the blockchain through 5G technology, in combination with the IoT (Internet of things) and IIoT (Industrial Internet of things). Innumerable applications of blockchain in the field of modern Industrial sciences are also discussed.

Acknowledgments

This book would not have been possible without the help and guidance received from various quarters. I would like to take this opportunity to express my sincere thanks and gratitude to all of them.

I am thankful to Almighty God for giving me this opportunity. I extend my deep sense of gratitude to my husband Dr. Y. Harold Robinson, my son H. Jubin, my father Mr. S. Eanoch, and my mother Mrs. E. Lizzy Pushpa Bai. Their moral support and encouragement at all stages have led to the successful completion of this book.

I extend my deep sense of gratitude also to my scholars and friends for contributing chapters to this book.

Finally, I would like to specially thank all those at CRC Press who extended their kind help, encouragement, and moral support.

Dr. E. Golden Julie

Each and every member of my family has played a significant role in shaping me into what I am today. God has blessed me with an adorable child and I extend my heartfelt thanks to my beloved son for his thoughtful trait. I thank my husband for his constant support in all my works. I am indebted to my parents, brothers, and sisters for motivating me and showering their prayers and blessing in all my activities.

Dr. J. Jesu Vedha Nayahi

We would like to acknowledge the help of all people and close reference research circle involved in this project, and more specifically, to the authors and reviewers who took part in the review process. Without their support, this book could not have become a reality.

We would like to thank each one of the authors for their contributions. Our sincere gratitude goes to the chapter's authors who contributed their time and expertise to this book. Second, the editors wish to acknowledge the valuable contributions of the reviewers regarding the improvement of quality, coherence, and content presentation of chapters. Most of the authors also served as referees; we highly appreciate their double task.

Finally, we appreciate the cooperation, support, and patience extended by our families for sacrificing family times to complete this project well in time.

Dr. Noor Zaman Jhanjhi

Editors

E. Golden Julie is currently working as Senior Assistant Professor in the Department of Computer Science and Engineering, Anna University, Regional Campus, Tirunelveli. She received her Ph.D. degree in information and communication engineering from Anna University, Chennai, in 2017. She did M.E. in computer science and engineering from Nandha Engineering College, Tamil Nadu, in 2008 and B.E. in computer science and engineering from Tamil Nadu College of Engineering, Coimbatore, Tamil Nadu. She has more than 12 years of experience in teaching. She has published more than 34 papers in various international journals and presented more than 20 papers in both National and International Conferences. She has contributed 10 chapters for Springer, IGI Global Publication. She is editor of the book *Successful Implementation and Deployment of IoT Projects in Smart Cities*, published by IGI Global under the series Advances in Environmental Engineering and Green Technologies. She is one of the editors of the book *Handbook of Research on Blockchain Technology: Trend and Technologies*, published by Elsevier. She is a reviewer of many journals like *Computers & Electrical Engineering* by Elsevier and *Wireless Personal Communication* by Springer Publication. She has given many guest lectures on various subjects such as multicore architecture, operating system, and compiler design in premier institutions. She is one of the recognized reviewers and translators for NPTEL Online (MOOC) Courses Certificate from NPTEL. She has been a jury in national- and international-level IEEE conferences, project fair, and symposium. She has completed nine NPTEL courses. She has got NPTEL Star certificate and Topper and Sliver certificates from NPTEL. She has attended Faculty Development Programmes to enhance the knowledge of student's community. Her research area includes wireless sensor ad-hoc networks, soft computing, blockchain, fuzzy logic, neural network, soft computing techniques, clustering, optimized techniques, IoT, and image processing. She is also an active lifetime member of Indian Society of Technical Education.

J. Jesu Vedha Nayahi completed her B.E. in computer science and engineering in 2001, M.E. in computer science and engineering in 2004 and Ph.D. in the field of privacy preserving data mining in 2016. She started her teaching career in 2001, and she was a university rank holder in her post graduation. She has published several papers in international journals and international conferences. Her research interests include data mining, machine learning, automata theory, and database concepts.

Noor Zaman Jhanjhi is currently working as Associate Professor with Taylor's University, Malaysia. He has great international exposure in academia, research, administration, and academic quality accreditation. He worked with ILMA University, and King Faisal University (KFU) for a decade. He has 20 years of teaching and administrative experience. He has an intensive background of academic quality accreditation in higher education besides scientific research activities. He has worked for a decade for academic accreditation and earned ABET accreditation twice for three programs at CCSIT, King Faisal University, Saudi Arabia. He also worked for National Commission for Academic Accreditation and Assessment (NCAAA), Education Evaluation Commission Higher Education Sector (EECHES), formerly NCAAA, Saudi Arabia, for institutional-level accreditation. He also worked for the National Computing Education Accreditation Council (NCEAC).

Dr. Noor Zaman Jhanjhi was awarded as top reviewer, 1% globally by WoS/ISI (Publons) recently for the year 2019. He has edited/authored more than 13 research books with international reputed publishers, earned several research grants, and has to his credit a great number of indexed research articles. He has supervised several postgraduate students, including masters and Ph.D. Dr. Noor Zaman Jhanjhi is an associate editor of IEEE ACCESS, moderator of IEEE TechRxiv, keynote speaker for several IEEE international conferences globally, external examiner/evaluator for Ph.D. and masters for several universities, guest editor of several reputed journals, member of the editorial board of several research journals, and active TPC member of reputed conferences around the globe.

Contributors

Sivaprasad Abirami
Department of Information Technology
Noorul Islam Centre for Higher
   Education
Thuckalay, Tamil Nadu, India

B. Akoramurthy
Department of Computer Science &
   Engineering
Velammal Institute of Technology
Chennai, Tamil Nadu, India

T. Ananth Kumar
Department of Computer Science &
   Engineering
IFET College of Engineering
Villupuram, Tamil Nadu, India

A. Anasuya Threse Innocent
Department of Computer Science &
   Engineering
Amrita School of Engineering, Amrita
   Vishwa Vidyapeetham
Bengaluru, Karnataka, India

A. Balamurugan
Sri Krishna College of Technology
Kovaipudur, Tamil Nadu, India

P. J. Beslin Pajila
Department of Computer Science and
   Engineering
Francis Xavier Engineering College
Tirunelveli, Tamil Nadu, India

S. R. Boselin Prabhu
Department of Electronics and
   Communication Engineering
Surya Engineering College
Tamil Nadu, India

Usharani Chelladurai
Department of CSE
UCE, BIT Campus, Anna University
Tiruchirappalli, Tamil Nadu, India

R. Dinesh Kumar
Department of Computer Science &
   Engineering
Siddhartha Institute of Technology and
   Science
Hyderabad, Telangana, India

Mabel Finney
Infoziant IT Solutions(P) Ltd.
Chennai, Tamil Nadu, India

S. Gomathi
Department of Computer Science and
   Engineering
Francis Xavier Engineering College
Tirunelveli, Tamil Nadu, India

S. Hemalatha
Sri Shakthi Institute of Engineering and
   Technology
Coimbatore, Tamil Nadu, India

T. Ilamparithi
Department of Computer Science
Manonmaniam Sundaranar University
   College-Puliangudi
Kanyakumari, Tamil Nadu, India

K. E. Kannammal
Sri Shakthi Institute of Engineering and
   Technology
Coimbatore, Tamil Nadu, India

M. Kavitha Margret
Sri Krishna College of Technology
Kovaipudur, Tamil Nadu, India

V. Kishore Kumar
Department of Electronics and
    Communication Engineering
IFET College of Engineering
Villupuram, Tamil Nadu, India

R. Malathi
Sri Shakthi Institute of Engineering and
    Technology
Coimbatore, Tamil Nadu, India

S. Matilda
Department of Computer Science &
    Engineering
IFET College of Engineering
Villupuram, Tamil Nadu, India

A. Mohana Priya
Sri Shakthi Institute of Engineering and
    Technology
Coimbatore, Tamil Nadu, India

S. Palanikumar
Department of Information Technology
Noorul Islam Centre for Higher
    Education
Thuckalay, Tamil Nadu, India

Seethalakshmi Pandian
Department of ECE
UCE, BIT Campus, Anna University
Tiruchirappalli, Tamil Nadu, India

M. Pavithra
Department of Computer Science &
    Engineering
IFET College of Engineering
Villupuram, Tamil Nadu, India

G. Prakash
Department of Computer Science &
    Engineering
Amrita School of Engineering, Amrita
    Vishwa Vidyapeetham
Bengaluru, Karnataka, India

R. L. Priya
Department of Computer Science and
    Engineering
Noorul Islam Centre for Higher
    Education, Kumaracoil, Tamil Nadu,
    India

A. S. Radhamani
Department of Electronics and
    Communication Engineering
V.V. College of Engineering
Tirunelveli, Tamil Nadu, India

R. Rajmohan
Department of Computer Science &
    Engineering
IFET College of Engineering
Villupuram, Tamil Nadu, India

S. G. Sandhya
Department of Computer Science &
    Engineering
IFET College of Engineering
Villupuram, Tamil Nadu, India

S. Sundaresan
Department of Electronics and
    Communication Engineering
SRM TRP Engineering College
Irungalur, Tamil Nadu, India

K. Suresh Kumar
Department of Electronics and
    Communication Engineering
IFET College of Engineering
Villupuram, Tamil Nadu, India

D. Vijayanandh
Hindusthan College of Engineering and
    Technology
Tamil Nadu, India

S. Vinila Jinny
Department of Computer Science and
    Engineering
Noorul Islam Centre for Higher
    Education, Kumaracoil, Tamil Nadu,
    India

1 Blockchain and Internet of Things

A Survey

P. J. Beslin Pajila
Francis Xavier Engineering College

E. Golden Julie
Anna University

S. Gomathi
Francis Xavier Engineering College

CONTENTS

1.1 INTRODUCTION

Blockchain is the most emerging technology that has come to be known from the statements "magic behind Bitcoin" [1], "Design the next OS for our society" [2], and "interrupt everything" [3]. The main goal of blockchain technology is to stop the communication through a middleman and give proper rights to the legitimate user [4]. Usually, the society trusts the mediator for the transfer of capital. The mediators protect the authentication of the users and record and validate the transfer details. The mediators are banks and governments. Blockchain technology performs the task of mediators with the help of composite cryptography architecture, thus replacing the mediators like banks and governments. Figure 1.1 shows the traditional design of a blockchain.

1

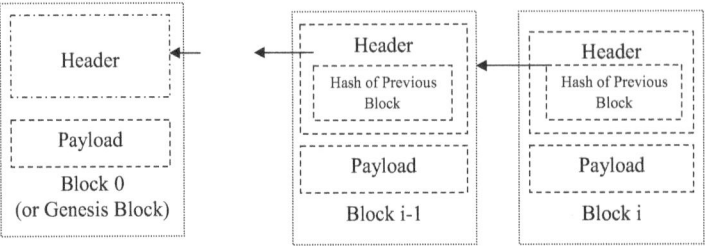

FIGURE 1.1   Traditional design of a blockchain.

According to computational concept, a blockchain is considered as a data structure, where data are stored inside blocks. Each block is linked with another block
in a sequential order. Blockchain is similar to the linked list concept; each element
is linked by using a pointer. Each block includes a payload and a header. Header
has information related to its length, content, 32 bit SHA256, i.e., the hash value of
the preceding block. Payload usually stores the activity details. The hash value of a
block depends on the preceding block. Transaction verification and validation form
into blocks, and storing data in a distributed ledger is the main goal of a blockchain.
Distributed harmony algorithm is used by the blockchain to find how and when to
include a set of transactions in the file.

All the new blocks can only be added to the blockchain if the bulk of the network
nodes agree to its inclusions, i.e., only if the blockchain users agree. Each network nodes
holds a local blockchain copy. It transmits over the network when a new block reaches
consensus. Numerous persistent copies can be created; if most of the nodes have similar copies of the blockchain, it is considered as more safe and loyal. Consensus process
is a major task, because if a node is added to a blockchain, it must be validated and
verified to check whether it is a legal transaction. It is also necessary to check whether
the added transaction is integrated with each other. All new transactions are included
in the blockchain, so verification of such transactions is the main target. If there is no
consensus process, then there is a possibility of various assaults [5].

The most important and the first consensus process is POW (Proof of Work),
which is handled by bitcoin. Most of the crptocurrencies adopt this mechanism.
Nodes (miners) in the POW process (mining) will fix complicated analytical puzzles
by using more computing power. There are two ways to solve complicated analytical
puzzles: the first one takes more time to determine the correct answer, and the second
one is simply guessing the answer. The first one is costlier, because solving puzzles
faster than usual requires faster calculating ability. While creating the blocks, it is
analyzed and created very slowly. If it is so, then it is easy to solve the puzzle. If the
blocks were created faster with no care, then it is very difficult to solve the puzzle [6].
We often add new blocks in the blockchain. While adding, it is necessary to attain
consensus so that it utilizes the computing power to solve the analytical puzzle.
It needs enormous quantity of energy and computing resources [7].

To overcome all those problems, the other consensus mechanism like proof of
stake (PoS) and proof of importance (PoI) are considered [8,9]. In Figure 1.1, the
new hash value relies on the preceding hash value in a block. So, if any new block is

inserted in the current blocks, it will generate a distinct hash value. This will lead to a descend effect for the successive blocks as well as the hash values. The consensus algorithm will drop the unwanted blocks in the blockchain. Consensus algorithm is very important for security purpose because the attacks will destroy, replicate, and modify the blocks in the blockchain. If the attack owns 50% of the blocks in a blockchain, it can administer the full blockchain.

1.2 INTERNET OF THINGS BASED ON BLOCKCHAIN

1.2.1 SMART ENERGY

Smart energy is transfer between prosumers and energy sellers. A smart meter will monitor the energy absorbed and sold for some period of time. Prosumer, smart meter, and merchant broadcast the information to the blockchain. Prosumer will communicate with other prosumers by using a public/private encryption key. It also uses this key to manage a secure or signed communication with the blockchain. There are four important entities: sellers, bidders (prosumers), smart meters, and smart contracts. The seller will start selling the energy by bargaining and advertising. Prosumer is used to purchase from many sources. Smart meter will record the energy spent at that particular period to check whether it is secure or not. Smart contract performs the operation and payment task. The working process is shown in Figure 1.2 [10].

The following are important actions of the smart energy process based on blockchain:

1. Smart contract installed in the blockchain
2. Seller bargains and advertises about the available energy
3. Powerful prosumer keeps track on the blockchain and acknowledges with bids
4. Smart contract uses the Vickrey auction to find out the winner and gives prize
5. The energy is given to prosumer with majority

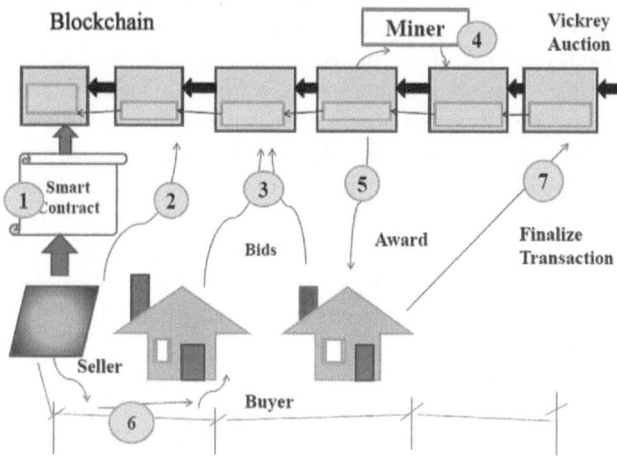

FIGURE 1.2 Blockchain of smart energy.

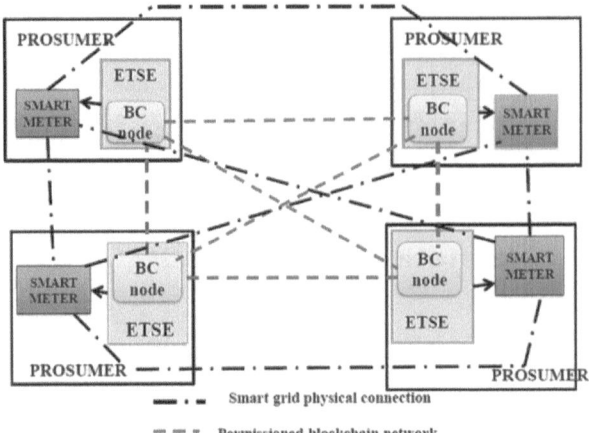

FIGURE 1.3   Physically connected prosumers.

6. Transaction starts

7. The transaction is decided and the amount is transferred

Smart meter is used to drop out the energy and currency after verification of the usage of energy by the prosumers [11]. Figure 1.3 shows four prosumers. Each prosumer has an Energy Trading and Security Enhancement (ETSE) software, a BC node, and a local smart meter. The software ETSE is installed inside each prosumer to collaborate between the local smart meters using an adapter. A blockchain node is deployed using ETSE in each prosumer. An interaction that was done through the BC node was secured with special functionalities, and it was mentioned using blue dotted lines. The three different layers of the prosumers premises (ETSE module) are shown in Figure 1.4. They are the Policy Management Layer (PML), the Energy Trading Layer (ETL) and the Security Enhancement Layer (SEL). PML is nothing but a set of rules. Using those rules the energy should be exchanged between the prosumers. Each prosumer has separate rules, and each one follows it separately. ETL manages and controls the smart meters from vulnerability. The local smart meters transfer energy based on contract. SEL is used for security purpose. It gets energy from the previous layer ETL. SEL verifies the energy; if it is vulnerable, it will violate it. Usually, SEL keeps on updating its security software if there is any solution to solve the particular vulnerability. It solves and stores it in a knowledge base [12]. Alternating Direction Method of Multipliers (ADMM) is an algorithm; it is also another method of smart energy consumption that decomposes a problem into subproblems, and it secures the transactions of energy naturally [13].

1.2.2   SMART CITY

Smart vehicle in the smart city has two main entities: Road side Infrastructure Units (RSI) and Service Providers (SP). Both of them are responsible for handling the blockchain. Each vehicle creates data that are gathered by other vehicles to verify traffic in

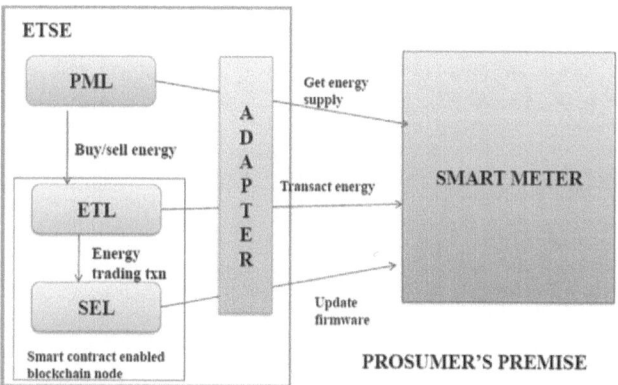

FIGURE 1.4 Framework of prosumers.

the city. Each vehicle is responsible for collecting data by using its sensors, and it is transferred to the nearby RSI. An advanced transaction is developed and is transferred to the immediate RSI. The blockchain has the block header where a public key resides, which is responsible for the verification of the advanced transaction. The existence of the vehicle must be verified so that it can get rid of Sybil attacks. The unknown vehicle should be validated with the location very carefully because it is used to collect information from other vehicles. RSI also verifies the vehicle once it sends any transaction. Each vehicle needs to register in the RSI when it needs to join the network [14]. The transfer of management information is known as transaction. Block Manager (BM) manages the blockchain. BM is responsible for the creation of transaction, authentication, and collection of separate blocks of transaction. Local BM is available to manage the Internet of Things (IoT) devices in the smart home. To circulate shared key among entities, Local BM (LBM) uses the Diffie–Hellman key distribution method [15]. In a smart city, all things are linked together and generate more data. In order to maintain the security of the data and IoT gadgets, blockchain technology is used [16].

1.2.3 SMART GREEN HOUSE

To safeguard crops from changes in Nature, many IoT gadgets are used. Blockchain is also available locally for security and privacy purposes. The blockchain concept is available through smart hub. The owner of the blockchain is responsible for fixing and eliminating a gadget. Each gadget in the IoT communicates with each other through a secured shared key where cryptography algorithm is used. The owner will handle all the transactions using the management header. The policies are available in the blocks of the blockchain. To eliminate the system overhead and delay all the nodes, assemble into a group called cluster. A cluster has a leader known as cluster head, which is determined by election. The delay and the overhead of the network is handled by the cluster leader. Data are stored globally in cloud. The data stored is also individual data, so it must be stored with a block number using the shared key it was encrypted. The shared key has originated from the crypto algorithm. The owner of

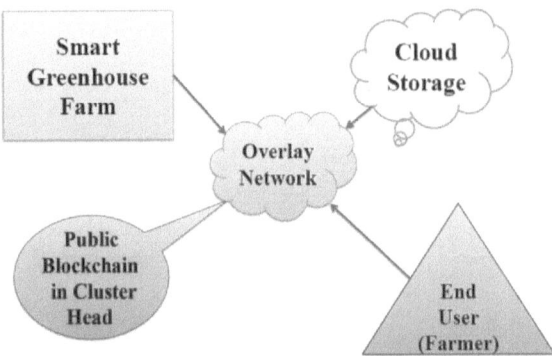

FIGURE 1.5   Structure of blockchain for smart green house.

the blockchain is referred to as the user. They are used to manage all the gadgets from anywhere through smart mobile phone and Internet. Figure 1.5 represents the structure of blockchain for the Smart Green House. IoT gadgets are heterogeneous, so various vulnerabilities of threats are possible in terms of availability, integrity, confidentiality, and authenticity [17].

1.2.4   ROBOTICS

Security is the main failure in the installation of robots for many applications. Blockchain plays an important role in providing security to defeat vulnerable attacks and threats. There are two different encryption methods: public key encryption and digital signature encryption. In public key encryption, each and every robot is encrypted using the public key. This key is accessible publicly in the network, but the private key is a secret one. The public key is shared among all the robots that need to interact with each other. Based on the private key, anyone can read the information. In digital key encryption, private key is used for encryption. Each robot encrypts the information using the private key; for other robots, those who want to read the data decrypt the information using the public key of the sender. It contributes authentication alone. According to public key cryptography, if the information is wrapped using blockchain technology, other robots can read the contents alone. The most outstanding example is linking ad hoc networks with the robot swarms that use the sensing applications that were distributed. The acceptance of blockchain technology in the robot swarms is more beneficial for the operators. They do not need to spend more time on training the new robots. Similarly, the new robots take less time for the learning process, because every transaction and agreement is reserved inside in the blockchain [18].

1.2.5   DATA PROTECTION

Drone is a device that collects information about soil moisture and composition using sensors and cameras. The collected information is transferred to the blockchain network to avoid the battery from draining and for securing the data. For each and

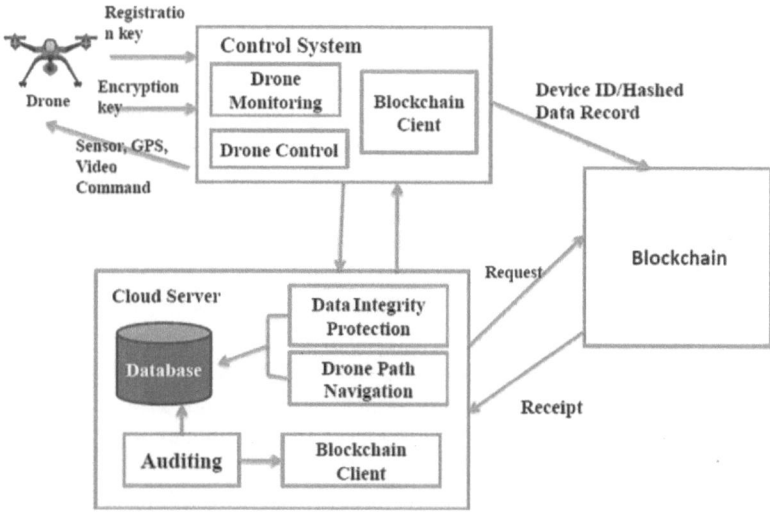

FIGURE 1.6 Drone system (data protection).

every data stored in the blockchain a receipt will be generated. Each drone has an ID. The data gathered are wrapped with the ID and are forwarded to the blockchain network so that the origin of the data can be found. The command system forwards a hash value for each data, which was already gathered in the blockchain network. The added record in a block is called as a transaction. The operation that took place should be stored in the blockchain network to find the vulnerability. As already known, each node in the blockchain network has awareness about the other blocks. All the data are protected. The drone-chain system's architecture has four important elements: drone, cloud database, blockchain network, and cloud server, as shown in the Figure 1.6.

Drone has a sensor that is used to collect data from soil structure and liquid content. The camera also helps to record the image and the video from the field. The control system collects the data from the drones. The collected data and the instructions together form a hashed data, and they are sent to the blockchain network for the protection. Cloud database deposit the data from the drone and the instructions from the control system and also the data access details. The data access details are mostly made using fingerprints. Cloud server performs the same function as the blockchain network [19].

1.2.6 TRANSPORTATION

Nowadays vehicles that are used to gather information from other vehicles using sensors are operated remotely. This significant development has created smart vehicles that are connected, which gives more important information to the manufactures, self-driving service cars, and insurance companies. The application that becomes more crucial is self-driving. The faulty part is identified after an accident based on the information available. To access the different data on the car, permissioned blockchain framework is used. To manage and control the data gathered, this framework is used, and it is combined with the VPKI (vehicular public key management) to accomplish privacy environment. To maintain a secure collaboration between the cooperating parties, four different nodes are used: leader, monitor units, validator, and client. There are many validators; one among them is chosen as the leader. The leader nominates a new block to the network based on the collected transactions. The validators spurt Byzantine compromise protocols to volatile malignant behavior of the participants and the leader. The validators spurt Byzantine compromise protocols to volatile malignant behavior of the participants and the leader. It creates a very big trust environment for all the participants. The hash values are compared to verify the honesty and purity of the data. The permitted user can access or control the transactions in the blockchain network. Since only the permitted user can access the transactions, identity is unique and that cannot be changed any more. To make the identity more secure and unique, the VPKI model that was recommended by the IEEE 1609.2 is used. Various identities that exist for no more than 5 minutes were given so that each and every vehicle could hold that identity for no more than 5 minutes. The transactions can be done on the blockchain with the vehicle with a valid alias identity [20].

Another method that secures the vehicles in transforming the information to each other is visual light and ultrasonic audio – a novel inter-vehicle collaborating system. The protocol used in these methods accepts TLS 1.2 handshake technique, because it exchanges the key on the Internet. TLS 1.2 protocol combined with the blockchain distributed hash table. It reduces the throughput specification, and the location of the vehicle is verified by using the visual light channel. Figure 1.7 represents the side-channel vehicle based on blockchain technology [21].

Here, blockchain is a ledger that protects the data that are available in the blockstack nodes. Each node is considered as a block that has separately distributed hash table with a public key for secure communication. There is a protected and more secure collaboration between the vehicles available in a platoon. The vehicles in a platoon have separate distributed public key inside a hash table. The above platoon system functions

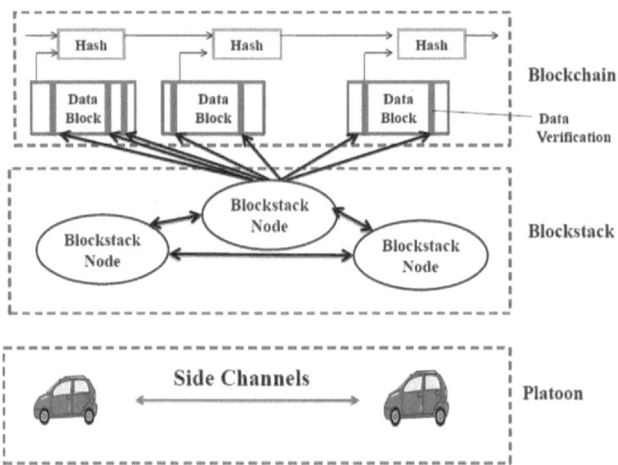

FIGURE 1.7   Side-channel vehicle based on blockchain.

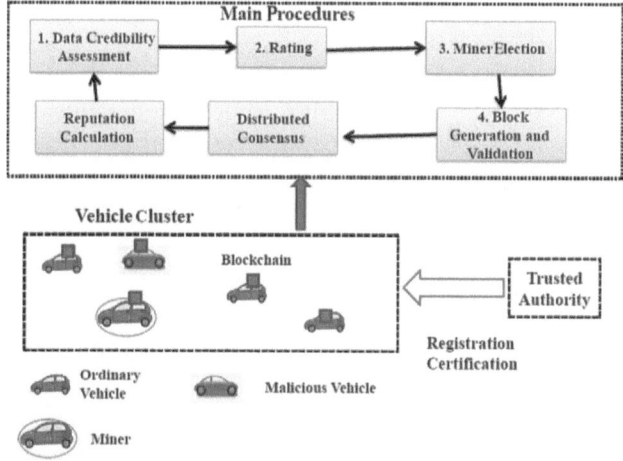

FIGURE 1.8 Blockchain-based reputation system.

with or without the Internet connection to the upper-layer blockstack [21]. The data credibility system is also used to discover the traffic environment based on the collected messages. These data are considered as the ratings of the vehicle. The ratings are packed inside a block, and all the blocks are connected with each other where the previous hash value was stored in the new one. A leader is selected from the vehicles based on the election process and that has the authority to send the rating blocks. The ratings are gathered inside the blockchain based on the ratings the vendor's message is evaluated. After the election, the winner bundle its ratings into the block [22] (Figure 1.8).

1.2.7 INDUSTRY IoT

BPI IoT is a new method, which is a blockchain platform of IoT; it combines an industrial IoT with the blockchain. It increases the performance of the CBM platform. Since it is combined with the blockchain, it adds a decentralized and peer-to-peer network. Smart contracts are installed in BPI IoT, which establishes an agreement between the manufacturing resources and the service customers. The main facilitator in the industrial machines is the IoT device. The IoT device permits the collaboration of the old gadgets with the cloud and blockchain network. It makes the machine to transfer the data to the cloud and messages to and from the smart contracts from the blockchain network. An IoT device has an Arduino and a Raspberry Pi. The Arduino includes a digital and an analog input and a digital output with sensors and actuators combined with it. A connection between the single-board computer (SBC) and the Arduino makes SBC to conquer the data from the Arduino and broadcast the signal to the actuators. The control drivers for the sensor and the actuator are mounted on the SBC. The device manager allows users to use a web interface to configure the SBC and also to monitor device status and statistics. The SBC's blockchain service connects with the blockchain network and sends/receives network transactions. IoT system has its own blockchain network account, and a blockchain wallet is maintained on the SBC [23].

The industry IoT based on blockchain can be elaborated in the following three aspects.

a. It provides an open manufacturing model for the future production environment focused on the link between customers and companies across the entire production ecosystem.
b. It offers a model of knowledge to explain the material and processes of manufacture information and service exchange.
c. It proposes an open manufacturing platform and process based on edge computing and blockchain [24].

1.3 BENEFITS OF BLOCKCHAIN

The following are the benefits of blockchain:

Empowered user: Consumers are in charge of their data and have no third party to trust.

Integrity: Users can be sure of executing transactions exactly as specified in the protocol. In order to ensure the reliability of data access and storage by IoT devices, information integrity must be maintained at all times. Once meaningful tools are provided to ensure data integrity, the blockchain helps. The validity of each block in the blockchain is checked by the hash value of each block, and the hash value of each block relies on the hash value of preceding blocks. Therefore, it not only guarantees the integrity of a new block, but it also applies the integrity test to all existing blocks. The integrity checking of the blockchain-based IoT systems also performs the same process.

Immutability and transparency: Transactions can hardly be changed on the blockchain. All transactions are available to all. The transactions are messages that are packed into a block and time-stamped. Each block is identified by the hash value known as cryptography hash. The address of the receiver, data payload, and the value are available in each transaction.

Enhanced interoperability of IoT systems: Blockchain technologies can add enhanced interoperability and privacy and security to IoT systems. In addition, blockchain can also improve IoT systems' efficiency and scalability. When IoT data are translated and stored in blockchains, the blockchain effectively boosts the interoperability of IoT systems. They convert, process, extract, compress, and finally store heterogeneous types of IoT data in blockchains.

Improved security of IoT systems: IoT data are stored inside the blockchain transactions, so they are considered more secure. The blockchain transactions are protected by using cryptographic public keys. However, the assimilation of IoT systems with the blockchain methodologies enhances the security of the IoT system by upgrading the firmware of IoT devices to fix insecure rupture.

Traceability and reliability of IoT data: There is no single point of failure capability to cope with malignant attacks. It is possible to identify and validate blockchain data anywhere and anytime. Meanwhile, all the historical transactions stored in the blockchains are traceable. In addition, the immutability of blockchains also ensures the reliability of IoT data, since any transactions stored in blockchains are almost impossible to alter or falsify. The blockchain incorporates many mechanisms such as public key encryption, private key encryption, authentication, and consensus and fault tolerance, which have been widely researched and tested for many networking scenarios in terms of security. Blockchain has therefore been described as a key technology for the design of safe and reliable IoT systems.

Autonomic interactions of IoT systems: Blockchain technology can automatically communicate with IoT devices or subsystems. Smart contracts enable Distributed Autonomous Corporations (DAC) to work automatically without human intervention, thus saving costs.

1.4 CONCLUSION

This chapter elaborates the various applications of IoT systems and how blockchain technologies are used to overcome the problems that were faced by IoT systems. The definition and working of the blockchain technologies were discussed in the introduction part. The benefits of blockchain technology after it was integrated with the IoT systems were discussed. The various architectures, methods, network structure, etc. of the IoT applications with the blockchain technologies were elaborated.

REFERENCES

1. K. Schulz. *Cryptocurrency 2015: The Rise and Adoption in E-Commerce*, 2015. URL: http://insights.wired.com/profiles/blogs/cryptocurrency-2015-the-rise-and-adoption-in-e-commerce.
2. R. Mulder, 2017. URL: https://medium.com/bitcoinevangelist/a-new-operating-system-for-our-society-77274c959037.
3. W. Long. *How Blockchain Technology Is Disrupting Everything*. URL: https://techdayhq.com/community/articles/how-blockchain-technology-is-disrupting-everything.
4. D. Bradburry. *How the Blockchain Could Change Your Life*, 2016. URL: https://www.thebalance.com/five-ways-that-the-blockchain-couldchange-your-life-391258.
5. F. Restuccia, S. D'Oro, S. S. Kanhere, T. Melodia, and S. K. Das. "Blockchain for the internet of things: Present and future." *IEEE Internet of Things Journal*, vol. 1, no. 1, 2018.
6. C. Dwork, and M. Naor, "Pricing via processing or combatting junk mail," in *Annual International Cryptology Conference*. Springer, 1992, pp. 139–147.
7. K. J. O'Dwyer, and D. Malone, "Bitcoin mining and its energy footprint," 2014.
8. A. Kiayias, I. Konstantinou, A. Russell, B. David, and R. Oliynykov, "A provably secure proof-of-stake blockchain protocol." *IACR Cryptology ePrint Archive*, vol. 2016, p. 889, 2016.
9. Z. Zheng, S. Xie, H. Dai, X. Chen, and H. Wang, "An overview of blockchain technology: Architecture, consensus, and future trends," in *2017 IEEE International Congress on Big Data (BigData Congress)*. IEEE, 2017, pp. 557–564.
10. A. Hahn, R. Singh, C.-C. Liu, and S. Chen, "Smart contract-based campus demonstration of decentralized transactive energy auctions," in *2017 IEEE Power & Energy Society Innovative Smart Grid Technologies Conference (ISGT)*. IEEE, 2017, pp. 1–5.
11. A. Laszka, A. Dubey, M. Walker, and D. Schmidt, "Providing privacy, safety, and security in IoT-based transactive energy systems using distributed ledgers," in *Proceedings of the Seventh International Conference on the Internet of Things*, ACM, 2017, p. 13.
12. F. Lombardi, L. Aniello, S. De Angelis, A. Margheri, and V. Sassone, "A blockchain-based infrastructure for reliable and cost-effective iotaided smart grids," in *Living in the Internet of Things: Cybersecurity of the IoT-2018*. IET, 2018, pp. 1–6.

13. E. Münsing, J. Mather, and S. Moura, "Blockchains for decentralized optimization of energy resources in microgrid networks," in *2017 IEEE Conference on Control Technology and Applications (CCTA)*. IEEE, 2017, pp. 2164–2171.

14. R. A. Michelin, A. Dorri, R. C. Lunardi, M. Steger, S. S. Kanhere, R. Jurdak, and A. F. Zorzo, "Speedychain: A framework for decoupling data from blockchain for smart cities," arXiv preprint arXiv:1807.01980, 2018.

15. A. Dorri, S. S. Kanhere, R. Jurdak, and P. Gauravaram, "LSB: A Lightweight Scalable BlockChain for IoT Security and Privacy," arXiv preprint arXiv:1712.02969, 2017.

16. S. Ibba, A. Pinna, M. Seu, and F. E. Pani, "Citysense: Blockchain oriented smart cities," in *Proceedings of the XP2017 Scientific Workshops*. ACM, 2017, p. 12.

17. A. S. Patil, B. A. Tama, Y. Park, and K.-H. Rhee, "A framework for blockchain based secure smart green house farming," in *Advances in Computer Science and Ubiquitous Computing*. Springer, 2017, pp. 1162–1167.

18. E. C. Ferrer, "The blockchain: A new framework for robotic swarm systems," arXiv preprint arXiv:1608.00695, 2016.

19. X. Liang, J. Zhao, S. Shetty, and D. Li, "Towards data assurance and resilience in IoT using blockchain," in *2017 IEEE on Military Communications Conference (MILCOM), MILCOM 2017*. IEEE, 2017, pp. 261–266.

20. M. Cebe, E. Erdin, K. Akkaya, H. Aksu, and S. Uluagac, "Block4forensic: An integrated lightweight blockchain framework for forensics applications of connected vehicles," arXiv preprint 5arXiv:1802.00561, 2018.

21. S. Rowan, M. Clear, M. Gerla, M. Huggard, and C. M. Goldrick, "Securing vehicle to vehicle communications using blockchain through visible light and acoustic side-channels," arXiv preprint arXiv:1704.02553, 2017.

22. Z. Yang, K. Zheng, K. Yang, and V. C. Leung, "A blockchain based reputation system for data credibility assessment in vehicular networks," in *2017 IEEE 28th Annual International Symposium on Personal, Indoor, and Mobile Radio Communications (PIMRC)*. IEEE, 2017, pp. 1–5.

23. A. Bahga and V. K. Madisetti, "Blockchain platform for industrial internet of things," *Journal of Software Engineering and Applications*, vol. 9, no. 10, p. 533, 2016.

24. Z. Li, W. Wang, G. Liu, L. Liu, J. He, and G. Huang, "Toward open manufacturing: A cross-enterprises knowledge and services exchange framework based on blockchain and edge computing," *Industrial Management & Data Systems*, vol. 118, no. 1, pp. 303–320, 2018.

2 Blockchain
A Review of Protocols and Standards

S. Gomathi
Francis Xavier Engineering College

Mabel Finney
Infoziant IT Solutions (P) Ltd.

P. J. Beslin Pajila
Francis Xavier Engineering College

CONTENTS

2.1   INTRODUCTION

Blockchain can be defined as a chain of the block that contains processed data. A blockchain is a ledger or a record of a transaction. It can be thought of as a chain or records stored in the form of blocks which are controlled by no single authority. Blockchain is basically a system used to keep data in the form of encrypted block. Blockchain first appeared in Nakamoto's bitcoin paper, which explains a new decentralized cryptocurrency (Nakamoto et al., 2008). In this technology, distributed consensus algorithm is used to solve traditional distributed database synchronized problems. It is an integrated solution and an unsettling innovation.

The blockchain records can be viewed as a log of ordered transactions. The blockchain technology is mainly used for secure transfer of items like property, contracts, etc. Here, a third-party intermediary like a bank or the government is not required. It is the backbone of the very famous cryptocurrency in the world – the bitcoin. In blockchain, a user can see every data, and he or she can control all his or her information and transactions. Our beloved prime minister, Mr. Modi, strongly believes that the blockchain concept will find new solutions in the agricultural sector. Transparent agriculture will be made with the help of real-time blockchain data.

Blockchains require data to be monitored in a digital manner. Since blockchain technology adopts new concepts, it has to cope up with existing technologies. For data authentication cryptographic algorithms are used in blockchain (Figure 2.1).

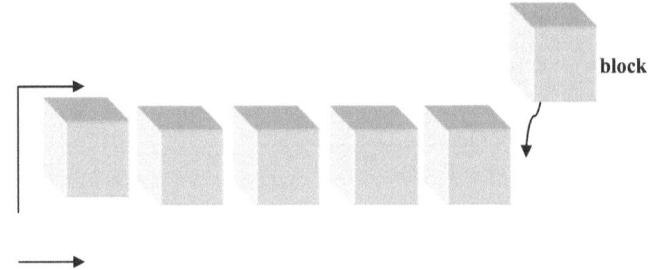

block

FIGURE 2.1   Blockchain.

2.1.1 Key Elements of Blockchain Technologies

The blockchain technology is composed of six key elements. They are discussed in the following sections.

2.1.1.1 Decentralized

The basic feature of blockchain is that data can be inserted and accessed on any node distributedly.

2.1.1.2 Transparent

The recorded data in a blockchain system is transparent to all the nodes.

2.1.1.3 Open Source

Most blockchain systems are open to all, accessible by all users.

2.1.1.4 Autonomy

Every node on the blockchain system can transfer or update data safely; the idea is to have trust from single person to the whole system, and no one can mediate it (Rokibul et al., 2010).

2.1.1.5 Immutable

The data entered in the system cannot be changed.

2.1.1.6 Anonymity

Data transaction can be anonymous in blockchain technologies.

2.1.1.7 Tamper-Proof

The data entered into the blockchain is tamper-proof.

2.1.1.8 User Empower

In the blockchain, a user can see every data, and he or she can control all his or her information and transactions.

2.1.1.9 Peer-to-Peer Network

Blockchain uses decentralized peer-to-peer protocol in which is each peer has the same capabilities.

2.1.1.10 Blocks

In blockchain, a number of blocks are connected to each other in a proper and linear sequential order. Data used for transactions get stored in files called blocks. Every block has a hash of the previous block (Figure 2.2).

2.1.2 Types of Blockchain

There are three basic types of blockchains: public blockchain, consortium blockchain, and private blockchain (Buterin, 2015).

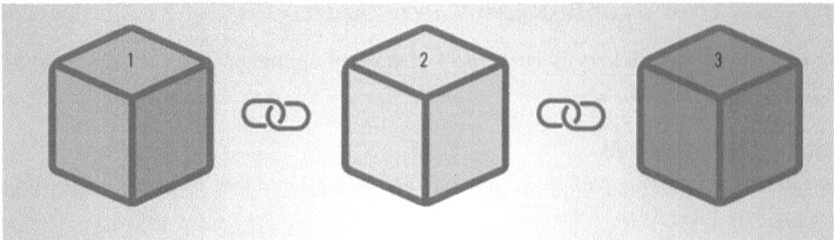

FIGURE 2.2   Blocks.

2.2   BLOCKCHAIN PROTOCOL

A blockchain protocol is a prevalent term used for agreements. These agreements are used for transaction validation within a blockchain network. Consensus protocol forms the basics of blockchain. It clearly specifies the working of blockchain. Usually, blockchain works on predefined rules which are agreed upon by all the participating nodes in the network.

These rules include the following:

- Rules for governing and validating transactions
- An algorithm that defines the mechanism for all participating nodes to interact with each other
- Application programming interface.

The rules that govern a blockchain network are referred to as blockchain protocols. They are essentially the common communication rules that the network plays by.

2.2.1   IMPORTANT FEATURES OF A BLOCKCHAIN PROTOCOL

- *Distributed ledgers*: Distributed ledgers are a collection of database records.
- *Smart contracts*: For transactions, embedded script can be used in the blockchain.
- *Consensus algorithm*: An algorithm that defines the way consensus will be reached on the network to verify the transactions.
- *Coins*: Every blockchain protocol needs a digital asset to keep the network running. These are also used as incentives for the peers who participate in the network. This entails the presence of digital assets such as coins and tokens. The two terms are often used interchangeably in the concept of blockchain but there is a difference between the two. Coins are defined at the lowest level by the protocol itself. Coins are the native digital asset of a blockchain network. For instance, bitcoin protocol's native currency is bitcoin.
- *Tokens*: Tokens are the digital assets that are defined at a higher level not by the protocol but by smart contracts. For instance, the Ethereum protocol has a native coin Ether.

- *Trustlessness*: The blockchain gives all peers an identical copy of each transaction, which eliminates trust, thus making a trustless, distributed network.
- *Decentralized*: A blockchain must be stored in a way that it can be accessed and copied on any node on the network.
- *Immutable*: All blockchain transactions are permanent transactions. Once a record is added, it cannot be changed.

2.2.2 DIFFERENT BLOCKCHAIN PROTOCOLS

2.2.2.1 Bitcoin

Bitcoin protocol supports crypto payment transactions over distributed network. Bitcoins are virtual currency, also called as cryptocurrency. Bitcoin includes cryptographic hash function and consensus algorithm. This protocol prevents duplication of payments and data fiction. Bitcoin prevents information breaches.

This transaction is done by exchanging digitally signed messages using bitcoin cryptocurrency software. Bitcoins are exchanged using the bitcoin protocol. Bitcoins are exchanged using bitcoin transactions (Barber et al., 2012).

2.2.2.1.1 Bitcoin Transactions

A bitcoin transaction requires a unique transaction ID, bitcoin addresses as the input, number of bitcoins to be transferred, and the receiver's bitcoin address.

2.2.2.1.2 Characteristics of Bitcoin Protocol

- Every node has access to complete information on the blockchain. Therefore, it is a decentralized one.
- Users can conduct nonreversible transactions without the need to explicitly trust a third party.

2.2.2.2 Ethereum

Ethereum is a public, open-source, blockchain-oriented protocol. In this protocol, a runtime environment is developed by programming language. The important component of this protocol is Ethereum Virtual Machine (EVM). Ether prevents unsolicited messages in the Internet. Ether and bitcoin have no of similar features. Bloomberg is a shared Ethereum software that is used by all users; but it is tamperproof. Cryptocurrency projects such as VeChain and OmiseGo were not launched using the Ethereum platform. It is a platform for dApp development and dApps such as Cryptokitties, Brave, and PundiX.

Ethereum blockchain has the same structure similar to bitcoin's. The entire transaction history is stored in every node on the network. Bitcoin uses an input in a new transaction to track the amount of the user's bitcoin. This transaction is known as unspent output or unspent transaction. Unspent outputs may be used to effect further transactions. Ethereum, on the other hand, uses accounts.

2.2.2.2.1 Characteristics of Ethereum Protocol

- This protocol enables developers to build and deploy distributed applications.
- It allows users to write their own applications.

2.2.2.3 Ripple

Ripple is an open-source payment protocol (Schwartz et al., 2014). Ripple acts as both a cryptocurrency and a digital payment network for financial transactions. Its native currency is termed as ripple. Ripple enables instant, safe, and almost free global financial transactions. The protocol supports tokens presenting cryptocurrency. Ripple is the biggest cryptocurrency in terms of market capitalization (Zhang and Lee, 2019).

2.2.2.3.1 Characteristics of Ripple Protocol
- Ripple protocol is known for its digital payment network.
- It has its own cryptocurrency, XRP.
- Ripple transactions are confirmed in seconds.
- Ripple transactions use less energy than bitcoin.

2.2.2.4 Hyperledger

Hyperledger is an open-source blockchain platform. It supports blockchain-based distributed ledgers and cross-industry blockchain technologies. Hyperledger frameworks are Hyperledger Burrow, Hyperledger Fabric, Hyperledger Iroha, and Hyperledger Sawtooth.

Hyperledger protocol mainly supports business transactions. Hyperledger is used to solve the problem of enterprise approval of blockchain. In Hyperledger, only trusted entities can join the network and verify the transactions.

2.2.2.4.1 Characteristics of Hyperledger Protocol
- Hyperledger blockchain technology is mostly used for business applications.
- Components such as consensus and membership services are allowed.
- It has a modular and versatile design.
- It preserves privacy.

2.2.2.5 R3's Corda

Corda is an open-source blockchain protocol used for industries such as financial services, insurance, healthcare, trade finance, and digital assets. This protocol is used for storing, coordinating, and controlling the financial agreements. This protocol captures the advantages of blockchain systems.

2.2.2.5.1 Characteristics of Corda
- Corda allows businesses to transact securely and seamlessly.
- This protocol organizes business operations.

2.2.2.6 Symbiont Distributed Ledger

Symbiont is used for institutional finance. It is a secure, high-performing Byzantine fault-tolerant distributed ledger. Symbiont can perform 80,000 transactions per second in distributed systems.

2.2.2.6.1 Characteristics of Symbiont Blockchain

Symbiont blockchain platform provides secure data distribution across financial institutions.

2.3 CONSENSUS ALGORITHMS IN BLOCKCHAIN

A consensus algorithm is a common agreement between all the peers in a blockchain network. Here, trust between all the nodes in a blockchain is achieved. The blockchain consensus protocol is responsible for controlling and coordinating the operations of all nodes in the consensus process.

2.3.1 Various Consensus Algorithms

2.3.1.1 Proof of Work (PoW)

In blockchain, new blocks enter an existing chain by using PoW consensus algorithm, which is used to confirm transactions and add new blocks to the chain. Information is shared among users by using digital tokens. Bitcoin uses this PoW consensus algorithm. It is also adopted by Ethereum, etc. (Figure 2.3).

2.3.1.2 Proof of Stake (PoS)

PoW is not scalable, so many projects replace PoW with PoS. Ethereum has moved to PoS consensus. PoS is a virtual agreement between the blockchain users. It validates blocks in the blockchain.

2.3.1.3 Delegated Proof of Stake (DPoS)

To avoid scalability issues, DPoS is proposed. DPoS is designed for ensuring secured transaction by voting and election process. It avoids malicious users. It is a fast, sustainable, and low power-consuming algorithm. This algorithm constructs stake methods for blockchain users.

2.3.1.4 Practical Byzantine Fault Tolerance (PBFT)

PBFT provides a practical Byzantine state machine which can even work in failures (Figure 2.4).

PBFT includes request, pre-prepare, prepare, commit, and reply phases. PBFT shows low algorithm complexity and complex practicality in distributed systems (Castro and Liskov, 1999) since it can achieve distributed consensus without carrying out complex mathematical computations (like in PoW). The transactions do not require multiple confirmations after they are finalized and agreed upon. PBFT suffered from communication overhead. Stellar is an advancement of PBFT (Zhang and Lee, 2019).

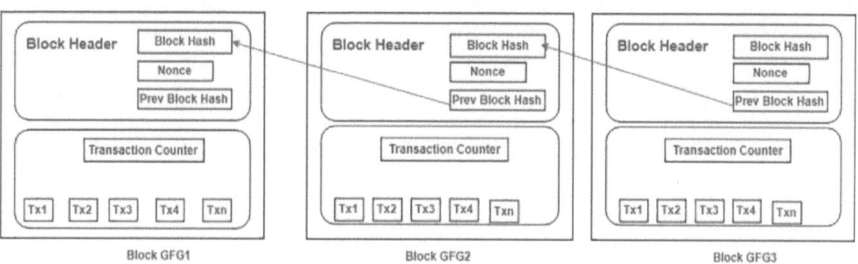

FIGURE 2.3 Proof of work.

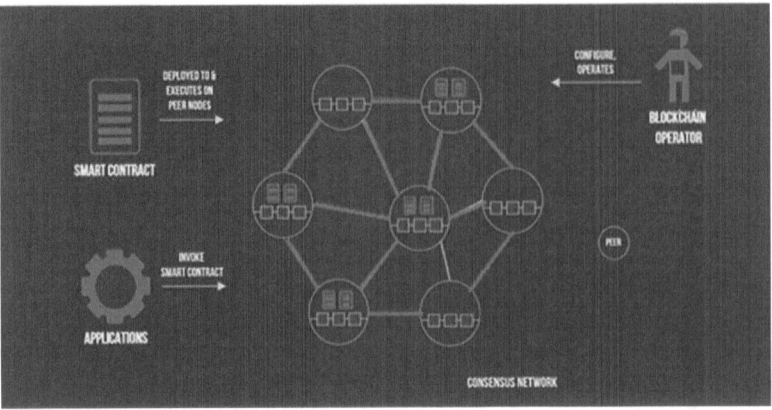

FIGURE 2.4 Practical Byzantine fault tolerance.

2.3.2 Consensus Algorithms

The main objective of consensus algorithms is effective decentralization. These networks are not only useful for financial applications but they also enhance the utilization on individual basis. It is a challenging task to use mining methods to introduce a new blockchain. From technical aspect, PoS is much more difficult to manipulate, and it protects against 51% of the attacks. For this reason, Ethereum has moved from a PoW to PoS. It is more difficult for malicious actors to attack the PoS network. PoS prevents entry of any unauthorized node in the network.

2.4 BLOCKCHAIN STANDARDS

The International Standards Organization (ISO) is developing standards for blockchain. The ISO plans to develop a terminology standard by the end of 2020. The standards are compliance, trust, security, integrity, and architecture.

2.4.1 Importance of Standards

The interoperable protocols such as Hyperledger Fabric, R3's Corda, and Ethereum differ in defining permissions from each other. Translating one to the other is risky. Consortium in the same industry affects the blockchain. Blockchain technology needs standard technical implementations, architectures, and standards.

2.4.2 Risk Factors

ISO identified blockchain industry-specific risk factors. Some of the highlights are as follows:

- Open-source projects won't care about standards.
- Industries can use their own standards.
- Blockchain fragments have to handle different problems.

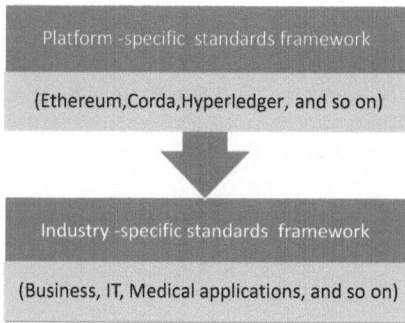

FIGURE 2.5 The types of blockchain standards framework.

2.4.3 BLOCKCHAIN STANDARDS FRAMEWORK

Blockchain standards are flexible with various technologies. They rely on ISO standards. Based on the blockchain usage, the blockchain standards framework is grouped as platform-specific standards framework and industry-specific standards framework. These standards may include distributed computing, security, governance, and business applications. Several other globally significant regulations impact blockchain. With different inputs, developing standards for blockchain is difficult (Figure 2.5).

2.5 CONCLUSION

This chapter covered the blockchain standards and blockchain protocols like Hyperledger Fabric, Symbiont, R3Corda, blockchain, Ripple, and Ethereum. In addition, this chapter reviewed and categorized the popular protocols used for networking, communication, and management. The chapter focused on analyzing consensus protocols, their feasibility, efficiency, and characteristics. A good consensus protocol must be developed by considering fault tolerance and proper application scenario. The chapter finally concluded that Hyperledger outperforms Ethereum and ripple. Ethereum is more flexible than node failures, but they are prone to security attacks. Ethereum requires more memory and disk usage than other protocols. Hyperledger uses low-level data model, but its elasticity enables modified optimization for queries.

REFERENCES

S. Barber, X. Boyen, E. Shi, E. Uzun. 2012. "Bitter to Better – how to make Bitcoin a better currency" (PDF). *Financial Cryptography and Data Security*. Springer Publishing.

V. Buterin. 2015. "On public and private blockchains", https://blog.ethereum.org/2015/08/07/on-public-and-private-blockchains/.

M. Castro, et al. 1999. Practical byzantine fault tolerance, *OSDI*, Vol. 99, pp. 173–186.

S. Nakamoto, et al. 2008. *Bitcoin: A Peer-to-Peer Electronic Cash System*. https://bitcoin.org/bitcoin.pdf

A. K. M. Rokibul, S. Tamura, S. Taniguchi, and T. Yanase. 2010. "An anonymous voting scheme based on confirmation numbers" *IEEJ Transactions on Electronics, Information and Systems* Vol. 130, No. 11, pp. 2065–2073.

D. Schwartz, et al. 2014. The ripple protocol consensus algorithm, *Ripple Labs Inc White Paper* Vol. 5, pp. 1–8.

S. Zhang, J.-H. Lee. 2019. Analysis of the main consensus protocols of blockchain, *ICT Express*.

3 Blockchain Technology
An Insight into Architecture, Use Cases, and Its Application with Industrial IoT and Big Data

K. Suresh Kumar and T. Ananth Kumar
IFET College of Engineering

A. S. Radhamani
V.V. College of Engineering

S. Sundaresan
SRM TRP Engineering College

CONTENTS

3.1   INTRODUCTION

Blockchain is defined as a chain of blocks that consists of specific information like data, images, or both. The report containing the database is secure, genuine, and grouped together in a distributed system (End-to-End network). It is a dispersed network where the computers are combined or linked to each other unlike the central server. It allows digital data to be distributed rather than copied. In order to provide transparency, security, and trust, distributed ledger is cast off. The blockchain containing data is recorded in a direct issue and the new square of the square chain comprises information from the previous squares. Blockchain technology can be applied in the banking sector, finance sector, manufacturing sector, professional services, energy and utilities, government and public goods, insurance, technology services, media, gaming, environment, generic, and healthcare [1]. The flourishing of the IoT (Internet of Things) upgrades the existing knowledge of things and also permits association of countless objects on Earth with the system. This shall interface automobiles, family unit apparatuses, and some electronics-based gadgets on a network. The framework enables distinguishing proof, area, following, checking, and comparing. Besides, IoT is the fundamental part of the IIoT (Industrial Internet of Things), which envisions conveying collected data and setting up industries by establishing a close relationship between customers and partners [2].

Information is a field having independent, deliberately separate data and managing them by using programming. In order to solve the problems related to security with personal information, blockchain is utilized in Big Data [3]. Due to the availability of massive data carrying personal information, it is not easy to handle it. The blockchain innovation is still in its initial stage and enormous applications scope is not available. Hence, it is predicted that owing to its trait of non-altering and its ability to recognize, the square chain normally has application and points of interest in ventures, for example, fund and credit detailing. This massive amount of data can be solved by using blockchain. Apart from handling, security can also be ensured by using blockchain. The most recent dynamic research has found that in the past two decades the use of distributed computing has become widespread. Driven by expansion in systems administration and dispersed designs, distributed (cloud) computing is an indication of conveyed frameworks that explored until the underlying origin of customer server design in 1958 [4]. Because of the quick development of distributed computing, it has been received as significantly effective in all sections of society, viz., the scholarly community, legislative establishments, and industry. Qualities of distributed computing, for example, were measured and dynamic access to the common groups of registering assets [5] were given, which empowered and acknowledged recent innovations and ideal models to satisfy the requests of rising applications that include logical, medicinal services, horticulture, keen city, and traffic board [6].

3.2 ARCHITECTURES FOR BLOCKCHAIN IMPLEMENTATION

Blockchain is one of the most advanced techniques used in recent times. The architectures employing blockchain technology are used mostly in financial industries. Presently, the square chain innovation is used for record-keeping, smart contracts, and digital notary, which has some features like decentralization, accountability, and security. The blockchain architectures can be understood by knowing about its main components and its interaction with outside environments. The built-on block architecture progresses as per the applications of blockchain.

The World Wide Web uses a traditional architecture that employs a client–server network. This technique improves the efficiency of operation and it is cost-effective. In the blockchain technology, the server stores all information in a centralized database where it can be easily updated and controlled by authorized administrators. The network is distributed so that every single participant who is in the blockchain network will maintain, update, and approve new entries. Data validity and security are ensured by each member in the block. The data within the blocks are encrypted using cryptography principles (e.g., steganography and few other algorithms). Figure 3.1 shows the architecture of blockchain.

3.2.1 INTEGRATION OF IoT AND BDA (BIG DATA ANALYTICS)

Ease-of-use system architecture is developed that integrates the blockchain, BDA, and IoT, which permits the retailers and sellers to check their supply chain frequently and enhance their sustainability most efficiently. The cost of implementation is analyzed and it provides a safe working environment [7]. The social sustainability of the supply chain is highly influenced by the blockchain technology. The integrated

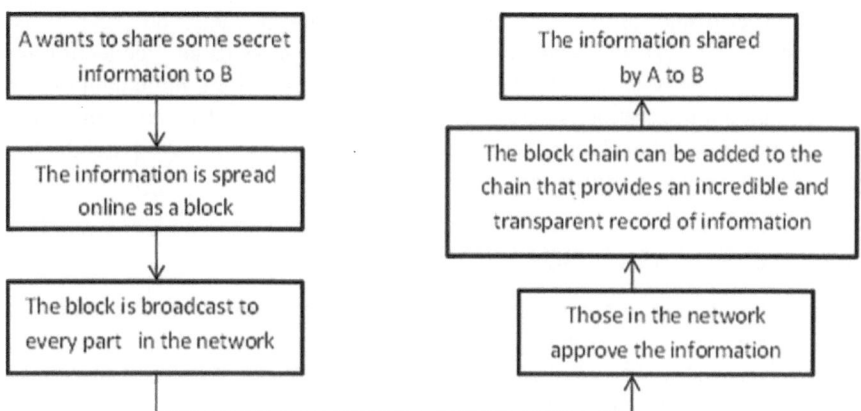

FIGURE 3.1 Example architecture of blockchain implementation.

architecture conducts an analysis, including the cost and challenges of implementation. This helps to understand the deployment of blockchain-based technology by enhancing the responsibility in the source chain. The monitoring and corrective actions are done by using IoT, which is made possible through the combination of IoT and BDA. Big Data will analyze the data about the machinery and devices and find the causes of failure. The IoT, in a joint effort for large information, assists organizations from being responsive to proactive [8].

3.2.2 BLOCKCHAIN ARCHITECTURES FOR PROJECT DELIVERABLES

The inventory chains in multiorganization ventures require the following innovations. An ongoing subsequent system is required for providing a global supply chain. For project-based business, distributed architectures are needed so that vendors could be engaged in tracking and tracing a project [9]. This architecture provides a pilot system that is used for controlling and measuring the supply chain processes that are traceable in real time. It is developed for integrating RFID, IoT, and blockchain technologies. The developed system is connected to the transport companies and suppliers for providing a sequence of unchallengeable businesses.

3.2.3 BLOCKCHAIN: A CLOUD MANUFACTURING SYSTEM

A dispersed system–based cloud-producing framework underpins the distributed system designs to improve the security and adaptability while working on cloud. The design comprises several layers – application, foundation, production, recognition, and asset. The idea of the engineering, including the protected information sharing, is examined and discussed with key advances that are required for the usage of the square chain design, which provides advantages in the form of security and scalability [10].

3.2.4 BLOCKCHAIN: INTELLIGENT TRANSPORTATION SYSTEM

For improving human work–life culture, electronic home appliances must be made advanced by using the IoT. This also includes enhanced safety and security. The application of blockchain, when combined with IoT, provides an architecture that can be used effectively in self-propelled businesses; it is helpful in maintaining the operational administration, safety, privacy, and traceability of the customer information that is very much confidential [11]. It consists of four-layered architecture: applications, gateways, processors, and sensors.

3.2.5 BLOCKCHAIN AND BDA FOR ENVIRONMENT

IoT aids the world by its applications like smart city implementations. In combination with the blockchain, it is used to construct a platform that is based on the functional flow of data and the structural composition of the blockchain in resource sharing and model exchange for a smart city [12].

The model developed helps compare carbon emissions with the green environment. The architecture consists of three layers: blockchain infrastructure layer, business application layer, and network layer. Alongside these layers, it is upheld by five help frameworks that can adequately share and realize the data assets. It additionally gives new plans to information proprietorship, peers, andexecutives, guidelines, and non-constant trade. Every association and offices have authority over the information tasks in the sharing of the administration data. Hence, the government information sharing has strong adaptability and is also highly reliable.

3.2.6 SMART HOME/IoT TOPOLOGICAL ARCHITECTURES BASED ON BLOCKCHAIN TECHNOLOGY

Even though various arrangements and stages have risen, this could be utilized for home-based robotization. A number of open sources that may lead in usage in organizations by way of time include technology-based arrangements. Figure 3.2 displays the protocol stack for home automation.

Hence, an IoT door can't be IP switch or some Access Points, albeit a few sellers may decide to bundle every one of these capacities in a physical gadget for comfort. In real business terms, IoT doors available at the present time give a legitimate total of sensor information streams, interpretation between sensor conventions whenever necessary, and information figuring capacities by handling the sensor information that is send before it moves to the examination framework either in the neighborhood or in the edge/cloud that conceivably bolster some of the cryptographic methods and approval/confirmation capacities [13].

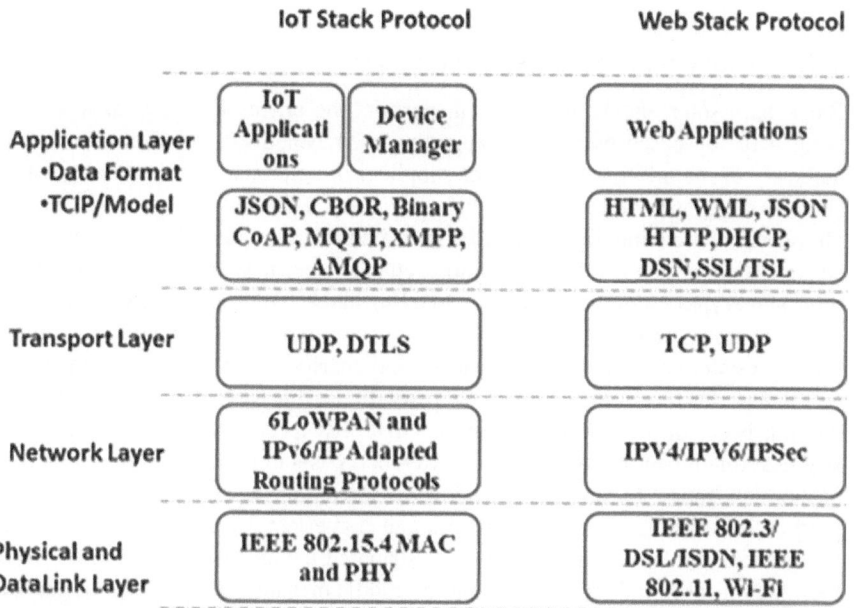

FIGURE 3.2 Protocol stack for home automation.

3.3 USE CASES FOR BLOCKCHAIN IMPLEMENTATION

The essential features of blockchain technology focus on settling issues that sway our lives. It is used in many cases that pertain to trust, security, accuracy, and transparency.

3.3.1 BLOCKCHAIN AS SUPPLY CHAIN

The usage of blockchain-based technology enables manufacturers to identify the sources or origins of certain goods and products through a supply chain management process. The misrepresented foodstuffs and the counterfeit items were pushed back toward a long way, which can increase both the potential of the consumers and their capacity to affirm the wellsprings of merchandise and items they purchase.

Ambrosius and Vechain are the blockchain technology-based platforms that target the consumers to identify the origin of food products by ensuring its safety and to confirm the quality and authenticity of the purchased products.

3.3.2 BLOCKCHAIN IN HEALTHCARE

In healthcare, blockchain provides transparency and integrity. Hence, it is needed much in the pharmaceutical industry to handle medical prescriptions and transportation of medical items and equipment. Also, blockchain paves way for transparency, accuracy, trust, and security of the data handling that are vital in medical field.

3.3.3 BLOCKCHAIN AND ENERGY MANAGEMENT

Blockchain solar energy management permits the clients to sell vitality segment lying with sun-based boards for expanding their benefit. It would also curtail the authority of the arbitrator and power suppliers in nature, where the vitality can be shared among each other in a straightforward way. The vitality segment is additionally worth referencing, for example, power record is an undertaking that employs a square chain, which permits clients to exchange electrical force with each other in return for crypto installments, thereby enabling blockchain solar energy management.

Apart from the industrial side, there are some other more applications, including real estate, music, politics, education, and charity. Some of the blockchain projects that perform certain technological revolution in businesses are discussed below. Aventus is one of the blockchain ventures that offer its clients the capacity to reclassify ticketing inventory network rules. Rentberry is an organization that uses block anchor innovation to pave the way toward getting an extended haul rental space progressively protected, straightforward, and cost-effective.

Helbiz uses blockchain innovation, distributed cloud/edge servers, and brilliant lock equipment to end the requirement for middle men and contributions of outsiders in vehicle rental procedures. Hemp Coin intends to help banking administrations to the cannabis business by using blockchain and digital forms of money.

3.3.4 BLOCKCHAIN'S IMPACT ON COUNTRIES

The misuse of the blockchain innovation that can yield huge rewards for countries accepting more technological change and countries that don't adopt the blockchain technology are much lagging behind. The country aided with blockchain technology for the last decade is Estonia. The data archives for governmental, computer-generated security, health, judicial, and finance purposes use this blockchain technology. Other nations that use this technology are Sweden, Switzerland, the United States, and South Korea.

3.4 IoT AND BLOCKCHAIN

The IoT is conquering the world not only in the form of home appliances but also in the field of industries and transport. Typically, for monitoring the traffic conditions, weather, and temperature, smart sensors are employed in each piece of the vehicle segments for viable working and this combined with blockchain gives an added advantage. IoT and blockchain integration is at the initial stage and recently more and more research is carried out to overcome the security issues in an effective way [11]. The combination of blockchain with the IoT permits a most authenticated way of transaction. Both are integrated for logistics management and automotive assets. The intelligent transportation systems used in car and other vehicles help the driver by warning about the collision and also monitor the driver's condition [15].

3.4.1 IoT-BASED SMART CITY APPLICATIONS USING BLOCKCHAIN

For smart city applications, the IoT-centered blockchain machinery is used for providing confidential information sharing. This uses multiple-edge gateways, which is accomplished by means of the multiple-path transmission strategy [15]. Here lightweight data consensus algorithm is used and hence it reduces the data theft by providing high data accuracy and reliability. The multiple-edge gateways are strictly steady, which reduces the network broadcast disruption. Thus, the energy loss is reduced and thereby delay in transmission is controlled. Figure 3.3 shows a blockchain network model for the small factory.

3.4.2 LIGHTWEIGHT DATA CONSENSUS ALGORITHM

The blockchain technology with IoT contains some layers like application, network, incentive, contract, consensus, and data, as presented in Figure 3.4.

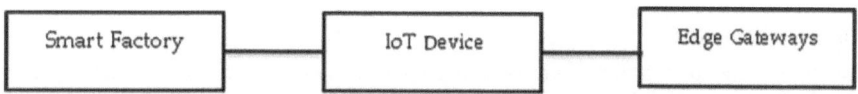

FIGURE 3.3 Example of network model for small scale factory.

Application Layer	Programmable Currency	Programmable Finance	Programmable Society
Contract layer	Script code	Algorithmic mechanism	Smart Contract
Incentive layer	Insurance mechanism	Distribution mechanism	
Consensus layer	PoW	PoS	DPoS
Network Layer	Peer-to-Peer Connection		
Data layer	Data block	Chain structure	Hash Function

FIGURE 3.4   Blockchain technology infrastructure.

An appropriate record exists as a database that remains collective and harmonized amongst IoT gadgets, and this record is kept in a decentralized framework. The record accounts for availing some exchanges concerning the whole gadgets in the system, for example, trading benefits or other information. Taking into account that the IoT gadget has constrained figuring power, stockpiling limit, and vitality assets, presently the circulated record is employed via edge passages, which have adequate assets in the system. The circulated record is kept by all edge portals in the manufacturing plant zone. The information transmitted through the IoT is recorded and the decentralized attributes guarantee that the information can't be altered.

Security and protection in correspondence with IoT gadgets have been looked into widely in later stages. Blockchain innovation, voices with other applications through IoT framework, and its productive execution have stood as a business experiencing exceptional examination. The decentralized control, unchanging nature, cryptographic safety, adaptation to internal failure, information trustworthiness, confirmation, and shrewd agreements of the blockchain are alluring highlights for the IoT [16]. The advancement of blockchain innovation and the pertinence of the blockchain stage with IoT paves way for a better environment, viz., bitcoin [17], as it is a decentralized payment system.

3.4.3   BLOCKCHAIN: IIoT

The use of IoT-based square chain innovation actualizes a compact, secure, and stage-skeptic square chain tokenizer. It depends on a mechanical information procurement unit that is ready to gather data from inheritance and present-day technologies that interfaces with the sensors. The information assimilated will be handled by empowering a modifying edge example and afterward sent to any blockchain stage [18].

3.4.4   IIoT AND BASIC IoT

3.4.4.1   Basic IoT

The interconnection among our ordinary gadgets (PCs, workstations, telephones, watches, and other handheld installed gadgets) is suggested by the IoT. Also, it suggests the interoperability among the abovesaid gadgets, just as the gadget independence, recognition, and situational mindfulness. An associated gadget fitted with sensors or actuators detects its general condition, senses what is going on, chooses brilliantly and autonomously, or speaks with different hubs or clients to make better choices. So, IoT is expected to add PC-based rationale to an object, which would later be observed or constrained by examination or motors [19]. The advancement of the IoT is a progressive fit to the necessities of mankind. It consolidates vehicles, medicinal services, wearables, retail, coordination, production, agribusiness, utilities, machines, and so on [20].

3.4.4.2 IIoT

The IIoT recommends the model or alleged utilizing process of robotization and information trade, which is present in social occasion industry. It joins businesses of the IoT, coordinates addition framework, and passes on enlisting [2]. The IIoT is used along with cyber-physical systems (CPS) for industries in order to provide and comprehend the digitized inventory showcase and fabrication, deals with its long-lasting accomplishments with advantageous, successful, and customized items [21,22]. The IIoT is a combination of numerous new advances, viz., independent machines, propelled mechanical autonomy, large information, cloud/edge processing, computerized universality, brilliant manufacturing plants, AI, and digital-physical based on IoT [23].

3.4.4.3 Blockchain Applications in IoT and IIoT

The blockchain is a continuous record of archives that are put away in a disseminated highlight point, which is autonomous of some focal power. Every database is encoded and stage stepped, by which the clients could reserve some options to get to and alter the squares intended for them if they have the private key [24]. Each square is associated with the past and the accompanying square with whole chain is re-energized each time a change is made [25]. Figure 3.5 shows the integration block diagram of IoT and IIoT using blockchain.

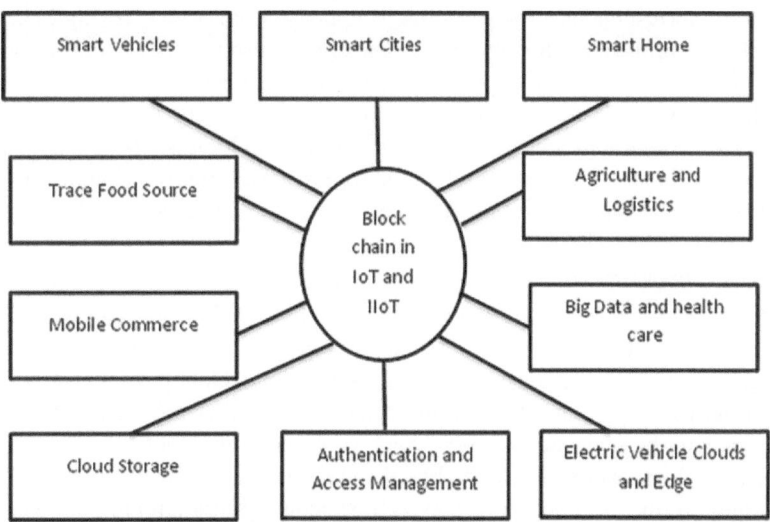

FIGURE 3.5 Application of IoT and IIoT using blockchain.

3.4.4.3.1   Smart Vehicles

Smart cities are a sort of constant improvement taking place as a generic development that gives a better quality life. Metropolitan compactness is the most noteworthy aspect of adroit urban territories. Owing to the enormous quantity of automobiles in these metropolitan territories, urban transportation blockage is continuously increasing. This makes the driver to find spots to leave the vehicle become unpredictable. According to the studies, we can make some clear conclusions that 30% of the transportation blockage is caused by drivers who search for parking spaces [26]. Considering this scenario, the drivers are in a great confusion that they do not know where to stop the vehicle as parking structures are unavailable. Another structure that arranges the IoT and a judicious model subject to troupe methodologies to streamline the figure of the openness of parking spaces is shrewd halting. Edge/cloud-driven IoT centered splendid transportation official's arrangement is made to ease the transportation and the time is improved practicality on the progress of the vehicles. The traffic inflow gauge changes the transportation improvement arrange time in a similar way and minimizes long holding up lines and blockages at convergences. The savvy route empowers the ideal appropriation of traffic to potential ways and thus improves street well-being at crossing points [27].

3.4.4.3.2   Smart Cities

Smart cities are planned to make the society and the government together by improving with specialist traits: splendid economy, sharp flexibility, wise condition, canny people, adroit living and keen organization [28]. IoT is another heterogeneous correspondence stage that gives universal access to assets and satisfies the client's needs. Creating a smart city worldview utilizes this correspondence stage for offering different types of assistance to the clients and conceding conveyed asset sharing [29]. A smart city plan gives innovative advances in private, business, administration and mechanical correspondence by providing simplicity of asset and sharing [30]. The physical layer procedures for improving security can use the features of IoT in the keen city. Consequently, multi-get to compact edge handling is used as a supportive unit that provides security in savvy urban zones. The correspondence properties are investigated for managing the safety problems in IoT-assisted splendid urban networks. Different strategies are reliant on resource assignment signal dealing and wiretap coding is examined for giving security in this city condition [31].

3.4.4.3.3   Smart Homes

The smart home is depicted as a pivotal use of omnipresent figuring that fuses insight into residences of the executives and their activities [32]. By utilizing blockchains, every gadget inside the home can demand information from the other interior gadgets to offer particular types of assistance. In the home, when anyone enters without authentication, the movement sensor, which is placed at the door side or any other area, directs the sensor signal to the board and the board attains its goal by switching its light to turn off condition. This will be achieved through program which is already encoded. The signal reaches the board and the controller checks whether the signal has been received from the right person using an acknowledgement process.

For this, the digger doles will avail the common key to another immediate special-ized gadget. When the key is obtained and checked the apparatus or machines will directly give access. The advantages of this device are that it is a two folded device. One fold is that diggers and the owners have a great overview of the systems that can share the common data. The next fold shows that that correspondence exists between the device and provides a guaranteed approach by providing a key which is typical [33]. The data or the information can be taken care locally and approved without dispersed capacity by using a shared key. To submit the key, the system should send deals to the digger and it has certain limits, which the excavator will make a shared permission key and it is then sent to the device and the key is set to aside. Then after getting the key, which is made close by storing the starting stage of the shared key available. By using a typical secret, the apparatus can store the data in the neighbour-hood accruing. After getting the screen trade, digger directs the present existing data of the referenced contraption to the supplicant. If the supplicant is permissible for getting the data or the information for a while, the excavator irregularly drives the data or the information until the person who is requesting the supplier who sends to the digger and repudiates the trade. The screen trades empower property holders by considering the watch cameras or various policies which can send periodic data [34].

3.4.4.3.4 Trace Food Source

The most needed application for blockchain is trace food source which is used to improve the IoT. The customary evolved way of life begins from producers, pro-viders, to merchants and this makes nourishment data befuddling and builds the trouble of following nourishment sources. Blockchain ensures that every exchange is time-stepped and carefully marked, such that it can be followed back to a pre-cise duration length and then beholding gathering is originated on the square chain through the open area. It is a consequence of the non-renouncement of the blockchain guaranteeing that somebody can't check the legitimacy of their mark scheduled on the documents or else that the creator's personality is the exchange they started, constructing the framework increasingly dependable. The change of record in world-wide status and the capacity of blockchain examined can give organization secu-rity and straightforwardness for every emphasis [35,36]. Blockchain guarantees the security of the store network and can be advantageous for taking care of emergency circumstances, for instance, item reviews in light of security breaks. The open acces-sibility of blockchains implies that every item could be followed to the wellspring of crude resources and the exchanges shall be connected in a sequence to clients distinguishing powerless IoT gadgets [24]. Presently, IBM and Walmart cooperate to make the natural pecking order straightforward. They consented to an arrangement with Tsinghua University to produce straightforward and productivity in inventory network record-maintaining. The Tsinghua University is a university that is working with Yonghui superstores. This analyses the fish production and records the same from the data obtained from the network in the store. Rather than normal analy-sis and physical investigation frameworks, blockchain gives an alternative exchange framework. Retailers could know about the providers who have been involved in exchange. Since exchanges are not put away in any single hub, it is practically dif-ficult to alter data without any problem. Simultaneously, it is simple for purchasers to get important data about merchandise, for example, industrial facility and preparing information, creation and termination period after checking the QR code utilizing a cell phone and reaching the administration in case of item disappointment. The administration will check the blockchains controlling important nourishment divisions [37]. Without blockchain, for the most part, it took almost 6 days, 18 hours and 26 minutes for Walmart for transporting mangoes to the first ranch. Presently, by way of the assistance of blockchain, purchasers need simply 2.2 second for acquiring the point by point data about products [38] completely.

3.4.4.3.5 Agriculture and Logistics

Agriculture and logistics are different in their platforms, as logistics are related to the smart building environment. Here, a logistics-based agriculture is being developed that consists of architecture for enhancing the production of the agricultural foods. The framework consists of enhanced intelligible architectural viewpoints and guidelines corresponding to it. The developed framework supports more than nineteen use cases which is used in the field of dairy production and farming, supply chains, fruits and vegetable production. The framework is used to model the cases in a punctual manner [39]. A systematic review is performed from 2002 to 2016 about the application of IoT in agriculture domain. Here it consumes a clustering-based IoT application that contains four domains like prediction, monitoring, logistics and controlling. The perception of key advances is utilized to build up the IoT applications [40]. The usage of recent knowledge and the apparatus that works in an IoT-based environment has created a smart solution that enhances the productivity in agriculture in combination with blockchain technology [41].

3.4.4.3.6 Mobile Commerce

Because of the blast of portable business (m-trade), information security issues are turning out to be increasingly significant and should be made tended. Blockchain implemented as a disseminated database suggested to tie down exchanges at versatile hubs to help shortest gadget-to-gadget m-trade information trade and distribution. An Android-based framework execution procedure has been presented in Ref. [42]. Blockchain ensures that the information trade between gadgets needn't bother with the association of any outsider. Some research fundamentally and evidently redefined communications amongst the networking units on a disseminated network [43]. The seller and the buyer make a contract at the initial stage that contains the real transaction information [44]. The two members affirm this agreement and distribute it in the blockchain framework. Accordingly, both get what they need safely with the assistance of the blockchain.

3.4.4.3.7 Cloud Storage

Distributed storage assumes a crucial job in the IoT improvement. One of the most familiar distributed storage is additionally a helpless connection in IIoT advancement. To start with, the IoT system considered with unified models of the cloud as a significant expense issue. IoT gadgets are associated, recognized, and confirmed by a server using cloud, where capacity and preparation are commonly accomplished. Regardless of whether the gadgets were a couple of bases separated; its association is unquestionably done through the Internet. The blockchain innovation, decentralization, could be accomplished, deprived of bringing together substances. Gadgets convey legitimately and trade dispersed information with one another, and naturally performing tasks through keen agreements. Next, each square of the IoT configuration can transform into a condition of dissatisfaction or a blockage which might hurt the whole framework. Designed for instance, IoT hubs remain defenceless against Domain name system DOS (Denial of Service) assaults and data robbery. Hoodlums could likewise harm frameworks and misuse information. On the off chance that the IoT gadget associated with a server is ruined, each hub associated with that server might be influenced. The blockchain confirms the legitimacy of the gadget character that scrambles and checks the exchange to guarantee that the message originator alone can send it, alongside the course of events in the chain, which assists clients with knowing the subtleties of the gadget or information chain plainly. Since the records are shared, a solitary purpose of disappointment record doesn't happen [45]. Third, concentrated cloud models are handily controlled. One can't guarantee that the data is utilized appropriately by gathering on-going information. The blockchain innovation permanence and decentralized access recognize and square malignant tasks. In the event that a blockchain modernizes for a gadget is debased, the framework discards it in Ref. [24].

3.4.4.3.8 Access Management and Authentication

Customary frameworks utilize an incorporated engineering and straightforward login, which is yet a risky for representatives and clients who are defenceless against taking or breaking passwords. Blockchain innovation gives solid verification and addresses single-point assaults, approving gadgets and clients under an appropriated open key framework, thereby supplanting customary passwords along with explicit Security Socket Layer (SSL) endorsements. Furthermore, the administration of endorsement information is made possible only on the blockchain, which could make it practically incomprehensible for an assailant to utilize a phony testament [36]. Blockchain character validation and access to the executive's methods can be utilized to upgrade IoT security. For instance, with this innovation, data about evidence of merchandise, personality, accreditations and computerized rights can be put away safely. Depending on the prerequisite, unique data that is entered is correct the blockchain might be unaltered. In IoT technologies, especially for the physical resources, the entities might solely collect the encrypted confusions of the system firmware in an isolated blockchain. The structure formed using a blockchain makes a constant database of system position and plan [24].

3.4.4.3.9 Electric Vehicle Clouds and Edge Computing

In this processing system, both vitality assets and data are traded, worked together and transferred between automobiles. The vehicles can be unconstrained system administrators, versatile information mini-computers or virtual force plants depending on various cases [2]. In security, the plot was worked by blockchain to set up conveyed agreement by means of information changes and criticalness changes dependent on hash tree figuring and timestamps, for example, confirmation of stake (PoS). The data and vitality exchanging record amongst the automobiles are blended and then that is made hooked on square chains in a straight back to back requesting, so the trade data couldn't be adjusted with no issues [46]. To fulfil the ever-expanding vitality asset requests of IIoT applications with developing quantities of gadgets, a confined blockchain-empowered secure vitality exchanging framework among vehicles was implicit [47]. A grouping of blockchain is proposed to achieve agreement forms surrounded by automobiles. It empowers conveyed accusing and releasing exchange safety and protection deprived of a confided outsider [48].

3.4.4.3.10 Big Data and Health Care

In the IIoT, human hereditary information is a higher priority than at any other time because of the canny creation. The business needs of blockchain money-related administrations are solid, bringing large information and related investigation instruments to the record given by the entire blockchain. As of late, a consortium of more than forty Japanese banks marked an agreement using a blockchain innovation which was considered an unidentified flow for utilizing the blockchain to encourage the exchange of assets concerning financial balances and thus accomplish on-going exchanges by the side of extremely ease. Customary constant exchanges are costly and a noteworthy motive is the latent peril factor. Twofold spending, where a comparable safety token is used double, remains the specific issue of constant exchange. The utilization of blockchain, to a great extent, maintains a strategic distance from this hazard. Massive information investigation accelerates the technique to distinguish buyer spending on designs and hazardous exchanges, quicker than at present. This decreases the asset or financial expense of constant transactions [49].

In IoT enterprises other than the financial business, the primary motorist of blockchain innovation is information security. The keen to well-being of retail and organization segments, organizations have started trying different things by means of blockchains to progress information to forestall hacking or information ruptures. In business, hereditary social data are a higher priority than at any other time owing to intellectual creation. The administration in India has built up a quality database framework dependent on square connections for 50 million individuals [50].

The medicinal services part is an issue-driven information and workforce serious area where the capacity to get, alteration and trust the information arising out of its exercises are basic for the tasks of the segment. It can be partitioned as the functions inside the social insurance area into triage, medical issue understanding, clinical dynamic, acknowledgment and evaluation of information based consideration. Figure 3.6 depicts the blockchain-based health care service. Accomplishing the ideal well-being results depends on connecting with a multidisciplinary group of comfort staff who can apply most proper information and advances and abilities when managing the patient. While working together with instructive establishments, the medicinal services area must give access to patients for preparing the understudies thereby creating and refining the essential aptitudes. Consequently, the instructive organizations furnish the area with qualified faculty. While teaming up, organizations should provide with an examination, designing motivation and well-being establishments must help with giving access to experts, witnesses, test people and perform tests.

While taking an interest in imminent clinical preliminaries, well-being foundations must help with creating, arranging, directing and detailing the examinations. Consequently, the examination and designing organizations furnish the social

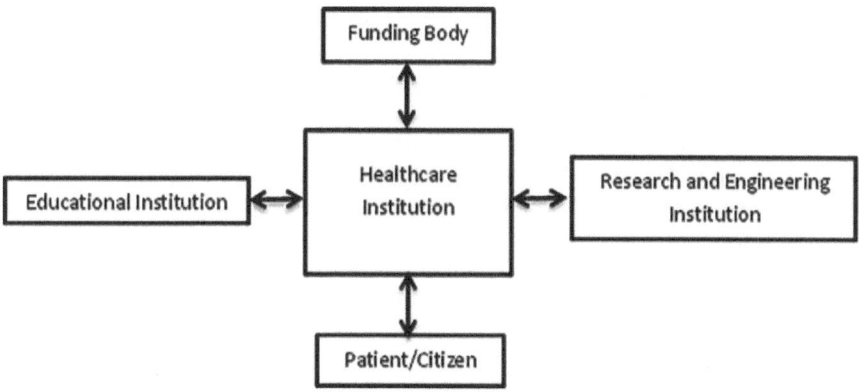

FIGURE 3.6   Blockchain-based big data health sector.

insurance area with refreshed information, methods and devices. Henceforth, the exercises of well-being organizations are firmly joined with foundations occupied with instructing safety workforce in biomedical research and designing. These activities require compelling trade of assents, tolerant related information and verifications and repayments forms, which successfully implies trading information across institutional fringes. Simultaneously, well-being foundations are commanded to ensure the profoundly touchy information that patients decide to impart to them [51].

3.4.4.4   Blockchain: IoT-Driven BDA

The development of technology provides some key benefits when combining blockchain and data analytics, along with the IoT. All three seems to have a certain difference but left with some relations which enhance the operations of the systems, its adaptability and practice. The underdeveloped technology that adopts blockchain and IoT is now producing a massive set of data and castoff across major industries like retailing, banking, healthcare, public administration, forecasting and implementation of smart cities along with public administration. All these were fully analyzed and interpreted using BDAs. Hence, the arrangement of these three brings massive changes and improvements [52].

3.4.4.5   Blockchain: Deep Learning and Artificial Intelligence (AI)

The AI plans to bring the IoT and Fog centers together for outstanding the burden condition and for persistently adjusting to give better Quality of Service, diminish authority utilization and enormous expense of the foundation. Human-made intelligence includes different pursuit calculations, AI, support learning and arranging [53]. AI gives a worthwhile road to streamline huge frameworks with gigantic measures of information with building effortlessness and effectiveness by permitting robotized dynamic rather than human encoded heuristics, which give increasingly effective choices rapidly. It can be utilized for preparing utilization information gathered from different IoT-based applications, for example, human services, agribusiness, brilliant home, and so forth [54].

Deep learning is another type of AI innovation that might be established around gaining from vast informational collections [55,56]. The centre of profound learning is to get a significant level of intelligent highlights from the crude information. Of late, deep learning has remained to fuel learning mechanisms that will help to understand the field of Deep Reinforcement Learning, which contributes trust while making improved models later on [57]. The convergence of deep learning plus distributed computing is making intriguing applications in both the fields are supplementing one another. How distributed computing is first talked about by supporting information researchers and later proceeds to how profound learning has been utilizing answers for different customary issues in distributed computing is discussed in Refs. [58–60].

3.4.4.6 Blockchain: Cloud Computing/Edge Computing

Distributed computing makes the product advancement procedure to be additionally effective. A few difficulties exist in consolidating programming designing and distributed computing, where these difficulties ought to be tended to information movement and identified as unique difficulties. Until cloud suppliers offer distinctive APIs for the contribution of cloud administrations, the programming and data ought to be moved to another cloud. Programming may be actualized and a couple of programming functions might be the legacy scheme [61]. For addressing this matter, while making and sending programming in fogs, silly conditions on unequivocal APIs ought to be sidestepped. Additional test is consistency and openness. In case the whole data is moved to fogs from close by and once the cloud gets attacked by any software engineers (Hackers) or affected by the astonishing calamity, it is difficult to recover the data. This might require the designers to get immediate neighbourhood reinforcement. Distributed computing gives additional opportunities to programme building analysts to investigate multilateral programming advancement [3]. A few analysts endeavoured to utilize distributed computing for diminishing the expense of activity, conveyance and programming advancement. In Ref. [62], creators investigated how to supplant learning. The board systems (LMS) organizes with a cloud arrange for sharing the data and joint exertion among school understudies. Programming structures are being displaced by cloud-based systems for saving cost and the most extraordinary use of assets. In late years, when data is developing into exponential rate, then it is hard to use the old ordinary technique for taking care of the information. The developing advancements dependent on IoT, blockchain, AI and computerized reasoning are opening another region of research in programme building, and the significant issue in these innovations is taking care of the tremendous sum and assortment of information. These inquire about additionally allowing the chance of new investigations and better approaches for taking care of the information in the cloud, and this outcome for the beginning of overhauled advancements like-Fog Computing-first utilized by Cisco so as to expand the present distributed computing foundation [63]. Programming organizations working for coding programmes build a reflection layer and offer help, called blockchain-as-a-Service [63]. All these rising zones are new, however, vigorously rely upon programming building.

It is a disseminated processing worldview, which performs a calculation on appropriated edge gadgets to empower the information assortment and correspondence over the system [64]. The edge figuring transfers the colossal size of data and information by taking care of edge-based devices instead of the cloud server, which expands the QoS (Quality of Service), decrease torpidity and transmission cost [65–67]. The time tricky applications might abuse from edge figuring; anyway, it then needs a steady Internet relationship with performing limits within the given time.

3.5   CONCLUSION AND SUMMARY

The blockchain is now enhancing its standards by encompassing in all fields of distributed computing. It is applicable in IoT, IIoT, Big Data, and cloud computing. Distributed computing is a rising worldview, empowering on request, metered access to figure assets (Process, Memory, Storage, and so forth) driving mechanical development and empowering topographically appropriated applications. For the improvement of distributed computing, the blockchain, IoT, IIoT, and Big Data are integrated. The architecture, use cases, and various applications using blockchain technologies are figured out by showing its ideal models and advancements that have been introduced and planned for its development. Further, the exploration zones identified with distributed computing have been recognized, talked about and the examination issues, difficulties are featured. An investigation analysis report is made that shows the impact of these rising standards (Blockchain, cloud computing, IoT, IIoT, and Big Data) on the advancement of distributed computing and thereby paving way for a modern era.

REFERENCES

1. https://mlsdev.com/blog/the-future-of-the-blockchain-technology-use-cases-geographical-expansion-potential-risks-and-challenges.
2. Q. Wang, et al. "Blockchain for the IoT and industrial IoT: A review." *Internet of Things* (2019): 100081.
3. J. Chen, Z. Lv, and H. Song. "Design of personnel big data management system based on blockchain." *Future Generation Computer Systems* 101(2019): 1122–1129.
4. M. J. Flynn. "Very high-speed computing systems." *Proceedings of the IEEE* 54 (12) (1966): 1901–1909.
5. B. P. Rimal, E. Choi, and I. Lumb. "A taxonomy and survey of cloud computing systems." in: *Fifth International Joint Conference on INC, IMS and IDC, NCM'09*, IEEE, 2009, pp. 44–51.
6. T. L. Casavant, and J. G. Kuhl. "A taxonomy of scheduling in general-purpose distributed computing systems." *IEEE Transactions on Software Engineering* 14 (2) (1988): 141–154.
7. V. G. Venkatesh, et al. "System architecture for blockchain based transparency of supply chain social sustainability." *Robotics and Computer-Integrated Manufacturing* 63 (2020): 101896.
8. https://www.experfy.com/blog/integrating-iot-with-big-data-a-revolutionary-step.
9. M. Sun, and J. Zhang. "Research on the application of block chain big data platform in the construction of new smart city for low carbon emission and green environment." *Computer Communications* 149 (2020): 332–342.
10. Z. Li, A.V. Barenji, and G.Q. Huang. "Toward a blockchain cloud manufacturing system as a peer to peer distributed network platform." *Robotics and Computer-Integrated Manufacturing* 54 (2018): 133–144.

11. P. Manjunath, R. Soman, and P. G. Shah. "IoT and block chain driven intelligent transportation system." *2018 Second International Conference on Green Computing and Internet of Things (ICGCIoT)*, 2018.

12. P. Helo, and A. H. M. Shamsuzzoha. "Real-time supply chain—a blockchain architecture for project deliveries." *Robotics and Computer-Integrated Manufacturing* 63 (2020): 101909.

13. D. Minoli. "Positioning of blockchain mechanisms in IOT-powered smart home systems: A gateway-based approach." *Internet of Things* (2019): 100147.

14. L. Suen, C. G. B. Mitchell, and S. Henderson. "Application of intelligent transportation systems to enhance vehicle safety for elderly and less able travelers." *Proceedings of the 16th International Technical Conference on Experimental Safety Vehicles*, Vol. 506, Washington DC, 1998.

15. W. Zhang, et al. "LDC: A lightweight dada consensus algorithm based on the block chain for the industrial internet of things for smart city applications." *Future Generation Computer Systems* 108 (2020, July): 574–582.

16. W. Yin, et al. "An anti-quantum transaction authentication approach in block chain." *IEEE Access* 6 (2018): 5393–5401.

17. I. Makhdoom, M. Abolhasan, and W. Ni. "Blockchain for IoT: The challenges and a way forward." *ICETE 2018-Proceedings of the 15th International Joint Conference on e-Business and Telecommunications*, 2018.

18. D. Mazzei, et al. "A Blockchain Tokenizer for industrial IoT trustless applications." *Future Generation Computer Systems* 105 (2020): 432–445.

19. D. Minoli, and B. Occhiogrosso. "Blockchain mechanisms for IoT security." *Internet of Things* 1 (2018): 1–13.

20. T. M. Fernández-Caramés, and P. Fraga-Lamas. "A review on the use of blockchain for the internet of things." *IEEE Access* 6 (2018): 32979–33001.

21. A. Rojko. "Industry 4.0 concept: Background and overview." *International Journal of Interactive Mobile Technologies (iJIM)* 11 (5) (2017): 77.

22. M. Zaouini. "Nine challenges of industry 4.0." http://iiot-world.com/connected-industry/nine-challenges-of-industry-4-0.

23. Unknown. "Industry 4.0: The fourth industrial revolution—guide to industrie 4.0." https://www.i-scoop.eu/industry-4-0/.

24. N. Kshetri. "Can blockchain strengthen the internet of things?" *It Professional* 19 (2017): 68–72.

25. H. Watanabe. *"Can Blockchain Protect Internet-of-Things?"* https://arxiv.org/ftp/arxiv/papers/1807/1807.06357.pdf. 2018.

26. S. C. K. Tekouabou, W. Cherif, and H. Silkan. "Improving parking availability prediction in smart cities with IoT and ensemble-based model." *Journal of King Saud University-Computer and Information Sciences* (2020, January 17): 1–11.

27. S. K. Sood. "Smart vehicular traffic management: An edge cloud centric IoT based framework." *Internet of Things* (2019): 100140.

28. D. Bruneo, et al. "An IoT service ecosystem for smart cities: The# smartme project." *Internet of Things* 5 (2019): 12–33.

29. D. Li, et al. "Improving communication precision of IoT through behavior-based learning in smart city environment." *Future Generation Computer Systems* 108 (2020, July): 512–520.

30. R. Morello, S. C. Mukhopadhyay, Z. Liu, D. Slomovitz, and S. R. Samantaray. "Advances on sensing technologies for smart cities and power grids: A review." *IEEE Sensors Journal*, 17 (23) (2017).

31. D. Wang, B. Bai, K. Lei, W. Zhao, Y. Yang, and Z. Han. "Enhancing information security via physical layer approaches in heterogeneous IoT with multiple access mobile edge computing in smart city." *IEEE Access*, 7 (2019): 54508–54521.

32. D. Mocrii, Y. Chen, and P. Musilek. "IoT-based smart homes: A review of system architecture, software, communications, privacy and security." *Internet of Things* 1 (2018): 81–98.

33. A. Dorri, et al. "Blockchain for IoT security and privacy: The case study of a smart home." in: *Proceedings of the IEEE International Conference on Pervasive Computing and Communications Workshops*, 2017.

34. M. A. Ferrag, M. Derdour, and M. Mukherjee. "Blockchain technologies for the internet of things: Research issues and challenges." *IEEE Internet of Things Journal* (2018), doi: 10.1109/JIOT.2018.2882794.

35. Y. Zhang, and J. Wen. "The IoT electric business model: Using blockchain technology for the internet of things." *Peer-to-Peer Networking and Applications* 10 (4) (2017): 983–994.

36. S. Ravindra. "The role of blockchain in cybersecurity." https://www.infosecurity-magazine.com/next-gen-infosec/Blockchain-cybersecurity/, January 8, 2018.

37. S. Ramamurthy. "Leveraging blockchain to improve food supply chain traceability." https://www.ibm.com/blogs/Blockchain/2016/11/leveraging-Blockchain-improve-food-supply-chain-traceability/ November 16, 2016.

38. S. Charlebois. "How blockchain technology could transform the food industry." https://theconversation.com/how-Blockchain-technology-could-transform-the-food-industry-89348. December 19, 2017.

39. C. Verdouw, et al. "Architecture framework of IoT-based food and farm systems: A multiple case study." *Computers and Electronics in Agriculture* 165 (2019): 104939.

40. J. M. Talavera, et al. "Review of IoT applications in agro-industrial and environmental fields." *Computers and Electronics in Agriculture* 142 (2017): 283–297.

41. J. Korczak, and K. Kijewska. "Smart Logistics in the development of Smart Cities." *Transportation Research Procedia* 39 (2019): 201–211.

42. Z. Li, et al. "Consortium blockchain for secure energy trading in industrial internet of things." *IEEE Transactions on Industrial Informatics* 14 (8) (2018): 3690–3700.

43. K. Suankaewmanee, et al. "Performance analysis and application of mobile blockchain." in: *Proceedings of the 2018 International Conference on Computing, Networking and Communications (ICNC)*, 2018, pp. 642–646.

44. K. Christidis, and M. Devetsikiotis, "Blockchains and smart contracts for the internet of things." *IEEE Access* 4 (2016): 2292–2303.

45. N. Joshi. "Distributed cloud storage with blockchain technology." https://www.allerin.com/blog/distributed-cloud-storage-with-Blockchain-technology. June 23, 2017.

46. S.K. Datta, et al. "Vehicles connected resources: Opportunities and challenges for the future." *IEEE Vehicular Technology Magazine* 12 (2) (2017): 26–35.

47. M. Conoscenti, A. Vetrò, and J.C. De Martin. "Blockchain for the internet of Things: A systematic literature review." in: *Proceedings of the 2016 IEEE/ACS 13th International Conference of Computer Systems and Applications (AICCSA)*, 2016, pp. 1–6.

48. J. Kang, et al. "Enabling localized peer-to-peer electricity trading among plug-in hybrid electric vehicles using consortium blockchains." *IEEE Transactions on Industrial Informatics* 13 (6) (2017): 3154–3164.

49. A. Venkat. "Introduction to blockchains & what it means to big data." https://www.kdnuggets.com/2017/09/introduction-blockchain-big-data.html.

50. T Ahram, et al. "Blockchain technology innovations." in: *Proceedings of the Technology & Engineering Management Conference*, IEEE, 2017, pp. 137–141.

51. A. Hasselgren, et al. "Blockchain in healthcare and health sciences–a scoping review." *International Journal of Medical Informatics* (2019): 104040.

52. P. Manjunath, and P. G. Shah. "IoT based food wastage management system." *2019 Third International Conference on I-SMAC (IoT in Social Mobile Analytics and Cloud) (I-SMAC)*, 2019, pp. 93–96.

53. S. J. Russell, and P. Norvig. *Artificial Intelligence: A Modern Approach*. Pearson Education Limited, Kuala Lumpur 2016.

54. S. S. Gill, et al. "Transformative effects of IoT, Blockchain and Artificial Intelligence on cloud computing: Evolution, vision, trends and open challenges." *Internet of Things* (2019): 100118.

55. A. A. Alli, and M. M. Alam. "SecOFF-FCIoT: Machine learning based secure offloading in Fog-Cloud of things for smart city applications." *Internet of Things* 7 (2019): 100070.

56. A. Dawoud, S. Shahristani, and C. Raun, "Deep learning and software-defined networks: Towards secure IoT architecture." *Internet of Things*, 3–4, 2018,

57. S. Tuli, N. Basumatary, and R. Buyya. "EdgeLens: Deep Learning based Object Detection in Integrated IoT, Fog and Cloud Computing Environments." in: *Proceedings of the 4th IEEE International Conference on Information Systems and Computer Networks* (ISCON 2019, IEEE Press, USA), Mathura, India, November 21–22, 2019.

58. K.K. Nguyen, D.T. Hoang, D. Niyato, P. Wang, D. Nguyen, and E. Dutkiewicz, "Cyberattack detection in mobile cloud computing: A deep learning approach." in: *2018 IEEE Wireless Communications and Networking Conference (WCNC)*, IEEE, 2018, pp. 1–6.

59. A.E. Eshratifar, A. Esmaili, and M. Pedram. BottleNet: A Deep Learning Architecture for Intelligent Mobile Cloud Computing Services. arXiv preprint arXiv:1902.01000. 2019.

60. S.A. Osia, A. S. Shamsabadi, A. Taheri, K. Katevas, S. Sajadmanesh, and H. R. Rabiee, Lane analytics. arXiv preprint arXiv:1703.02952.

61. S. Yau, and H. An. "Software engineering meets services and cloud computing." *Computer* 44 (10) (2011): 47–53.

62. T. Østerlie. *Cloud Computing: Impact on Software Engineering Research and Practice*. Norwegian University of Science and Technology (NTNU), Trondheim, 2009.

63. I. Scirlet. Cloud Technology in the era of IoT, Blockchain, Machine Learning and AI, White paper, available at https://blog.usejournal.com/cloud-technology-in-the-era-of-iot-blockchain-machine-learning-and-ai-4f1a19476b32. 2018.

64. S. Singh, and I. Chana. "QoS-aware autonomic resource management in cloud computing: A systematic review." *ACM Computing Surveys (CSUR)* 48 (3) (2016): 42.

65. F. Bonomi, R. Milito, J. Zhu, and S. Addepalli. "Fog computing and its role in the internet of things." in: *Proceedings of the first edition of the MCC workshop on Mobile cloud computing*, ACM, 2012, pp. 13–16.

66. A. A. Alli, and M. M. Alam. "SecOFF-FCIoT: Machine learning based secure offloading in Fog-Cloud of things for smart city applications." *Internet of Things* 7 (2019): 100070.

67. A. Dawoud, S. Shahristani, and C. Raun. "Deep learning and software-defined networks: Towards secure IoT architecture." *Internet of Things* 3–4 (2018): 82–89.

4 IoT and Blockchain

A. Mohana Priya, R. Malathi,
S. Hemalatha, and K. E. Kannammal
Department of Computer Science and Engineering,
Sri Shakthi Institute of Engineering and Technology

CONTENTS

4.1 INTRODUCTION

Though Internet of Things (IoT) and blockchain are emerging in their own way, the convergence of these two technologies will help leverage each other. Though the analytical capability of IoT had paved new paths in many enterprise applications, it has also led to serious security issues. Blockchain has proved itself to be a strong and secure technology with the success of bitcoins. IDC states that 20% of all IoT deployments will enable blockchain-based solutions by 2019. Using blockchain for IoT deployments will definitely open up a new era.

This chapter focuses on the following topics:

- How IoT works?
- What are the issues in IoT?
- What are some of the related technologies of IoT?
- How blockchain can help overcome the some of the issues in IoT?
- What are the design considerations to be kept in mind while designing the distributed ledger-based IoT architecture?

The recent trends in computing technologies have brought much more changes in all the fields and made effortless usability with hidden technical exertions. IoT is one of the emerging platforms where the things communicate with each other without or less human intervention. It makes things (embedded, nonphone devices) [1,2] smart with the presence of Internet; it mainly makes use of sensors, actuators, wireless

FIGURE 4.1 IoT in day-to-day Life.

sensor networks (WSN), smart cameras, radio frequency identification (RFID) readers, and tags. IoT has made humankind dependent on automations in many fields, including industrial automations, agriculture, and healthcare. IoT is [3] about the smart technology – remote monitoring, control, and the application of these technologies. Rather than using IoT as a single technology, a combination of many cutting edge technologies such as artificial intelligence, machine learning, and cloud computing are used in modern applications.

Figure 4.1 explains that IoT plays a vital role in all fields, from home to business, including healthcare, agriculture, traffic analysis, etc. IoT needs Internet in the forms of LAN, ZigBee or 6lowPAN, and things involving wireless sensors, actuators, and RFID devices that can produce results in a meaningful way.

4.1.1 IoT ARCHITECTURE

The IoT architecture comprises of four components (Figure 4.2). They are as follows:

1. **Sensor nodes:**
   A sensor node is a component that can work as an individual device or as a group of nodes together. It helps to receive signals from the environment that can be converted to human understandable data. Other than sensor nodes, some of the IoT devices include RFID tags and readers, thermal and smart cameras, accelerometers, and GPS.
2. **Gateway/sink:**
   The raw data that are collected by the sensor nodes are moved to the processing unit storage through the gateway. Here, the gateways can be a wireless medium (Wi-Fi, ZigBee) or a wired network. It acts as an interface between the nodes and the server.

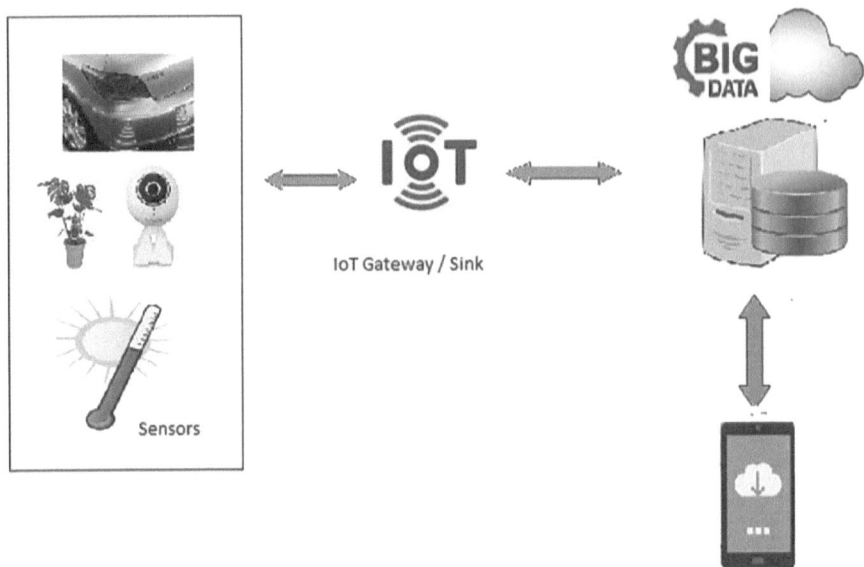

FIGURE 4.2   IoT architecture.

3. **Server:**

   The server receives the data, preprocesses, and filters and produces the useful information. It is stored in the server and also gives result to the user interface. According to the application, the server may itself control the sensor nodes by giving appropriate solution or the information would be passed on to the user interface.

4. **User interface:**

   The user interface can be a mobile app or any PC where the collected information can be viewed or used for further applications.

For instance, a moisture sensor senses the moisture level of the land and communicates with the server through the gateway on a regular basis. The raw data are processed and stored as information to be accessed by the end user, who will eventually view the information and water the land if the reported moisture level is below the required level.

   IoT has its own unique characteristics [6].

1. **Intelligence:**

   IoT is a combination of software and hardware, algorithms, and computations, which makes it smart. Ambient intelligence enhances its capability by enabling things to respond to a particular situation in an intelligent manner and supports them in performing specific tasks. Among various smart technologies, IoT is considered to be a very intelligent technology in relation to interaction between devices. By providing standard input methods and graphical user interface, the interaction between the user and the device is achieved.

2. **Connectivity:**

Connectivity plays a vital role in IoT by connecting objects together. It is considered as an essential part in IoT because object-level interactions lead to collective intelligence. It provides network accessibility and compatibility in things. By this connectivity, the networking of smart things and applications will create new market opportunities for the IoT.

3. **Dynamic nature:**

The key activity of IoT is to collect data from its environment; this is achieved with the dynamic changes that take place around the devices. The state of these devices changes dynamically, for example, sleeping to waking up and vice versa, connected and/or disconnected as well as the context of devices, including, location, temperature, and speed. In addition to the state of the device, the number of devices also changes dynamically with person, place, and time.

4. **Enormous scale:**

The number of devices that need to be managed and communicated with each other will be much larger than the devices connected to the current Internet. The management of data generated from these devices and their interpretation for application purposes becomes more critical. Gartner (2015) confirms the enormous scale of IoT in the estimated report, where it stated that 5.5 million new things will get connected every day and 6.4 billion connected things will be in use worldwide in 2016, which is up by 30% from 2015. The report also forecasts that the number of connected devices will reach 20.8 billion by 2020.

5. **Sensing:**

Sensors play a major role in measuring or detecting any changes in the environment. It provides the means of developing capabilities that represent a clear understanding of the physical world and its people.

6. **Heterogeneity:**

Heterogeneity in IoT is one of the key characteristics. Interactions between devices in IoT have different hardware and service platforms, and they also communicate with different networks. So, it's the duty of IoT architecture to support direct network connectivity between heterogeneous networks. Interoperability, information transparency, technical assistance, and decentralized decisions are considered to be heterogeneous things designed in IoT based on their environment.

7. **Security:**

IoT devices are naturally vulnerable to security threats. As we gain efficiencies, novel experiences, and other benefits from the IoT, it would be a mistake to forget about security concerns associated with it. There are high levels of transparency and privacy issues with IoT. It is important to secure the endpoints, the networks, and the data that are transferred across all means, creating a security paradigm.

There are a wide variety of technologies that are associated with IoT that facilitate in its successful functioning. IoT technologies possess the abovementioned characteristics, which create value and support human activities; they further enhance the capabilities of the IoT network by mutual cooperation and become part of the total system.

4.1.2   APPLICATIONS IN IoT [15]

The applications of IoT are expanding from its own root, making them the competitor of themselves. IoT has replaced much human-involved work, reducing the burden on us. For instance, the smart refrigerator intimates the milk vendor to deliver milk when the tray is empty; the smart irrigation system reduces the farmers' concern of the farming land by automating the irrigation process based on the moisture level, soil type, etc. Many governments, including Oman (Duqm smart city), have started smart city proposals with IoT and related technologies after South Korea's first smart city, SongDo [4].

We can't even imagine how much everyday life has been influenced by IoT. IoT is everywhere, from our smartphones to our home appliances. All that can connect and be managed through the Internet via IoT [7].

4.1.2.1   Wearable Gadgets

When you buy an IoT gadget that you can wear, it looks a bit cool. Some popular gadgets, such as Fitbit and smart watches, make our life so easy; many people first check their Fitbit to determine which meal to eat. These devices are suitable for people with high calories and blood pressure problem.

4.1.2.2   Smart Home Appliances

Nowadays PC gadgets have IoT as well as the home apparatuses. Imagine a scenario where I reveal to you that the machines that you use in your everyday life at your homes, for example, TV, refrigerator, AC, heater, light switches, entryway locks, vault locks, and so on have IoT support. Big technology companies, such as Samsung and LG have provided appliances to the world that can be controlled from anywhere in the world with the assistance of your cell phone through IoT. You have seen numerous commercials in which they show that you can turn on and off you AC from your office or school.

4.1.2.3   Smart City

Many developed nations such as the United States, Japan, and Norway have a savvy city, so what a smart city. A smart city is where everything is overseen appropriately with the assistance of the IoT idea. From water supply to executives to traffic lights, everything is controlled by IoT, and with the assistance of your cell phone and the web you can make up for the lost time with all the fixings. With the help of IoT, many information can be recorded, so that the information can be inferred and analyses for enhancement of the future.

4.1.2.4 Amazon Go

Amazon is one of the largest tech organization of the world. Amazon has established a retail shop where you can shop anything once you are an Amazon client. Amazon Go is not like other typical retail shops; it is a cutting edge shop with a lot of sensors. In Amazon Go, you don't need to go to the registration counter for making payment. You can simply pick the thing you need to purchase and leave the shop. The minute you leave the shop, the sum will be deducted from your Amazon wallet. It is that straightforward.

4.1.2.5 Smart Grids

Earlier, when electricity was made available to the consumer, power used to stream in just a single way, and the provider couldn't get constant information about the usage of power. The grids with the assistance of IoT take care of this issue now, and the customer can communicate with the power supplier and the supplier can supply the power in a mechanized manner.

4.1.2.6 Smart Farming

With the rise of many environmental concerns, such as climate change, environmental has a major impact on agriculture and we simply cannot ignore the farming culture. In farming nations such as Japan, smart farming has led to better productivity and efficiency and considerable results. Nonetheless, there is indeed a lot more to be done in smart agriculture, and IoT plays a major role in this.

4.1.2.7 Smart Supply Chain

We also order our stuff online; we get a feature that allows us to keep track of our order and can be able to check the order until it reaches our location. Life becomes easy and satisfactory with this smart supply chain.

Figure 4.3 explains the above mentioned IoT applications.

4.2 APPLICATION IMPLEMENTATION

In this section, we see two different implementations of IoT.

4.2.1 Garden Monitoring System and Controlling Pesticide Using IoT [16]

Currently, monitoring operations are done manually. Many people develop garden, but they do not maintain it due to their work engagements and surroundings. People leave their garden without growing any plants. Plants in the garden should be protected from any infections due to light, insects, deficiency, or any other factor. Pests in plants lead to irregular crop yield, which is not desirable. Nowadays, diseases in plants are common despite of pest control. These diseases are identified based on visible symptoms such as colored spots on leaves and streaks on leaves. As the disease spreads, the visible symptoms also increase. With machine-controlled technology

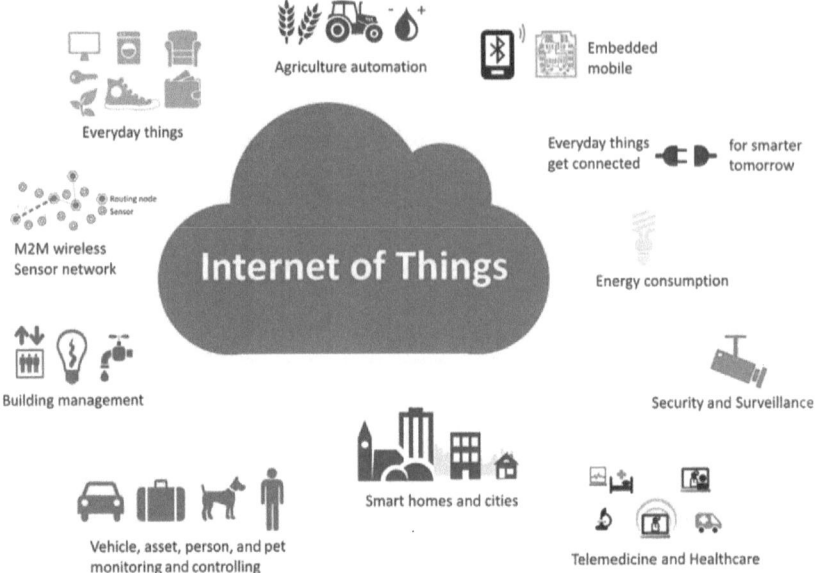

FIGURE 4.3 IoT applications. (Courtesy: [7].)

of irrigation, human intervention can be minimized. Whenever there is a change in water content of soil, the moisture sensor senses the change and gives an interrupt signal to the app. This moisture's data, i.e., the notifications, will be sent on the cloud using IoT, which can be accessed by the app. The system can unceasingly send the info on the cloud. The person can control the irrigation system through the Android App. In India, almost every farmer uses insecticides and pesticides to protect his crops from diseases and pests. They interpret weather on the basis of their experience, and when they detect pest on crop, they spray pesticides to protect their crop from disease and pest attack. Although these chemicals are saving their crop, but soil fertility is decreasing day by day. Inhaling these chemicals may lead to liver disorder, asthma, cancer, etc. Agriculture is one of the most ancient activities of man, in which innovation and technology are usually accepted with difficulty, only if real and immediate solutions are found for specific problems or for improving production and quality.

4.2.1.1   Sensors with Raspberry Pi

Figure 4.4 shows the devices that convert the electrical signals into digital signals. They are known as sensors. The different types of sensors incorporated in this system are listed next.

1. Humidity sensor is used to measure the humidity content of the soil.
2. Temperature device is employed to measure the temperature of the soil.
3 Moisture sensing element is employed to measure the wet content of the soil.

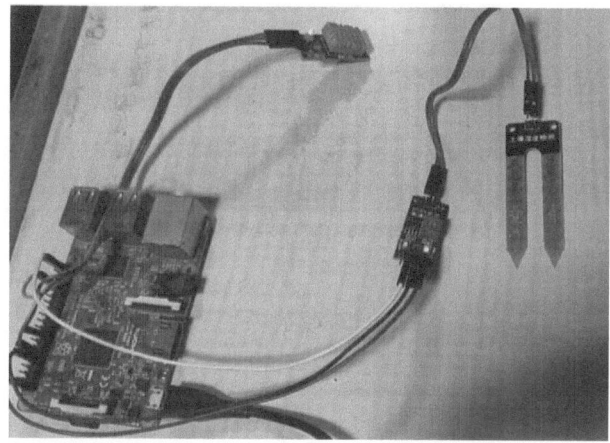

FIGURE 4.4   Sensor connectivity.

4.2.1.2   Block Diagram

Figure 4.5 shows a block diagram for sensor connectivity, where different sensors are connected with Raspberry pi. It has a mobile app where live data are stored in the firebase.

Figure 4.6 shows a block diagram for image processing.

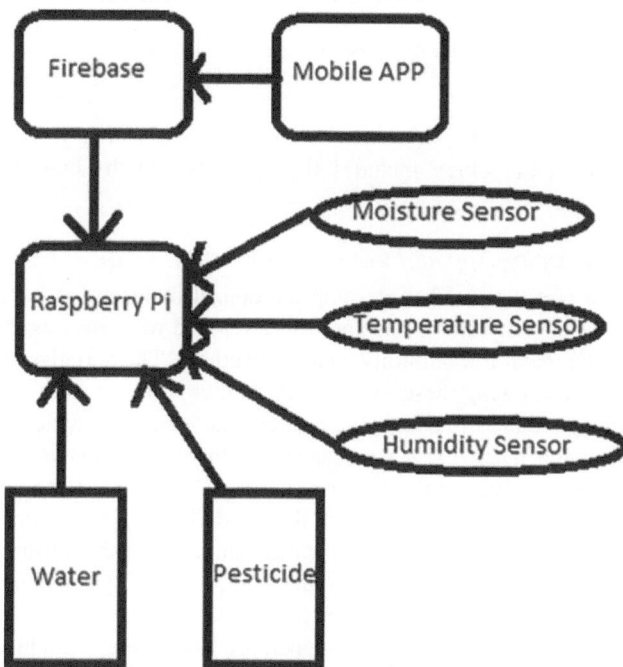

FIGURE 4.5   Block diagram for sensor connectivity.

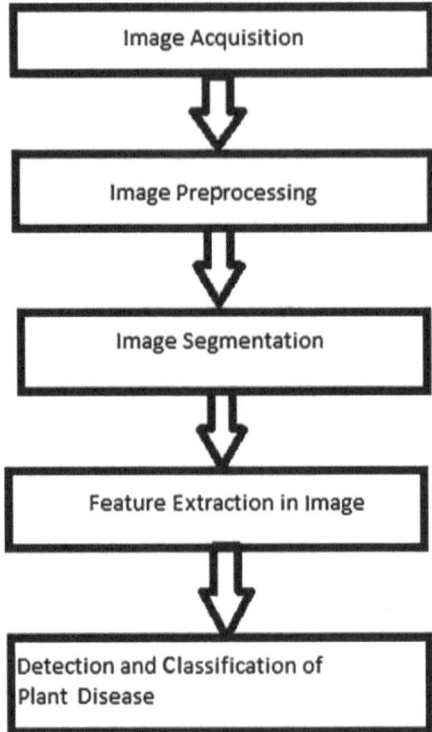

FIGURE 4.6 Block diagram for image processing.

4.2.1.3 Implementation

Implementation methodology includes the modules, which show how they are implemented.

4.2.1.3.1 *Developing App and Connecting It with Firebase*

Android studio is a tool used to develop the app. It includes the features such as button, toggle, and many more. The app was developed with features that are useful for noting the temperature, humidity, and soil wetness. The app gives the values of these features; by knowing these values, they may provide any alert message to the user. By noticing the alert message, the user can automatically switch on the button that is set in the app. Then, the person can supply the water to plants.

It consists of varied events for automating the good Garden. This mobile application permits the user to observe and manage the good Garden System domestically or remotely. Whenever the sensor crosses the maximum or threshold value, it alerts the user by push notifications and allows to take control of the system from the remote location. This methodology is cost-effective.

Figure 4.7 shows that when an app generates values, they should be recorded. The live data for the soil should be noted for proper growth of the plants; and monitoring should be regular. These values are used in alert messages, and automatically the

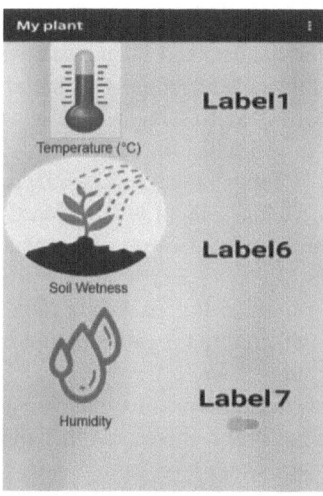

FIGURE 4.7 Layout of mobile app.

water supply to plants is done. Firebase could be a mobile and net app development platform in hand by Google. It helps to build better mobile applications.

It provides functionalities such as analytics, databases, messaging, and crash reporting. It is engineered on Google infrastructure and scales mechanically. It is easy to integrate firebase with android and the web. APIs are packaged into single SDK; hence it can be expanded to more platforms. It provides a real-time database and backend service. The real-time values from the sensors are uploaded to firebase through Raspberry pi. Figure 4.8 shows that the base is integrated with mobile apps for management purpose.

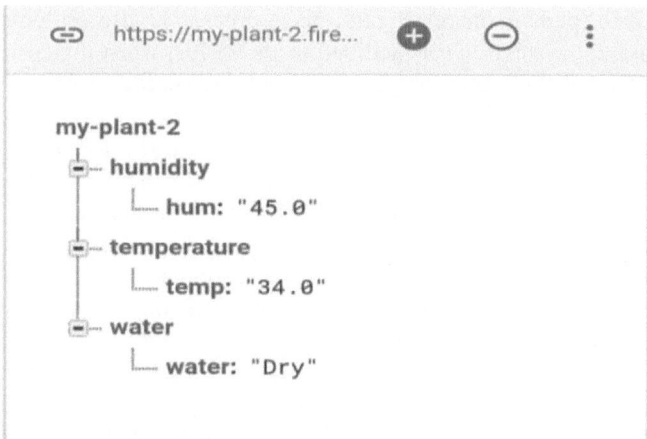

FIGURE 4.8 Sample firebase data stored.

4.2.1.3.2   Sensor Connectivity with Raspberry Pi

The capacitance of a capacitor that uses soil as an insulator depends on the soil water content. When connecting this capacitor (made of metal plates or rods imbedded within the soil) beside Associate in Nursing with a generator to create Associate in Nursing circuit, changes in soil moisture can be detected by changes within the circuit operational frequency. Temperature sensor can measure temperature from −55 °C to +150 °C [5]. The voltage output will increase 10 mV per degree rise in temperature. The humidity sensor SYS-1 is used for sensing the humidity. It delivers instrumentation quality RH (Relative Humidity) sensing performance during a low-price, solder ready SIP (single in-line package). Relative humidity is the actual amount of vapor in the air compared to the total amount of vapor that could be held in the air at a given temperature.

The system designed does not only monitor factors such as moisture, humidity, temperature, and ultrasonic but it also monitors other parameters according to the requirement, for example, if the water level in tank is reduced to a minimum, then the motor switch is turned on automatically and remains so until the water level of the tank reaches the maximum level.

The detector technology enables the system to elevate to the next level, which helps the users to utilize their investment in an economic manner. If soil nutrient sensors will be put in, then the system can be modified to supply fertilizers to the garden precisely. This system saves hands and expeditiously utilizes the water resources offered, ultimately resulting in a lot of profit. The feedback provided by the system can improve the implementation of the agriculture method.

4.2.1.3.3   Image Processing for Disease Identification

Image process is the use of PC algorithms to perform image process on digital pictures. As a subcategory or field of digital signal process, digital image processing has many advantages over analog image processing. It permits wider algorithms to be applied to the computer file and might avoid issues like the build-up of noise and signal distortion throughout a process. Since picture area units outline over two dimensions (perhaps more), digital image process is also sculptured within three-dimensional systems. That modified in the 1970s, when digital image processing proliferated as cheaper computers and dedicated hardware became available (Figure 4.9).

Images then might be processed in real time, for some dedicated problems such as television standards conversion. As general-purpose computers became faster, they began to take over the role of dedicated hardware for specialized and computer-intensive operations. With quick computers and signal processors offered in the 2000s, digital image processing has become the most common form of image processing and is generally used because it is not solely the foremost versatile methodology, however additionally the most cost-effective. In order to maintain the plants, initially an android app was developed, and the values were stored in firebase, by sensing these values from the environment. Image processing is done to identify the diseases in plants. In tomato plant, yellow curl leaves disease has been seen, which is

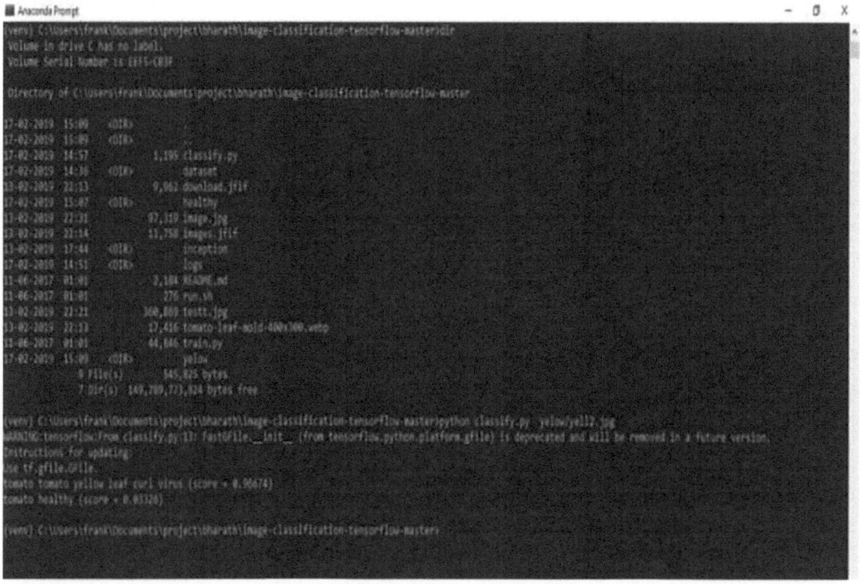

FIGURE 4.9 Image processing values for tomato curl plant.

caused by yellow curl virus and the values of healthy plant and infected plant values are displayed.

The system reduces human intervention in farming and becomes critical in creating more productive procedures for farming-related exercises. In the proposed work, median filtering is favored over manual channel, smoothing filter, and oscillation threshold channel because single outliners are separated in it. For identification of the infected leaves, CNN approach gradient descendant calculation is utilized. This gives more precise outcomes for a particular disease.

4.2.2 IMPLEMENTATION OF SMART HOME AUTOMATION USING ARDUINO

Smart home automation helps one to control the room lights, fans, or any electrical equipment at his or her convenience using his or her smartphone either using Bluetooth or Internet. Any device can be controlled on fingertips. There are many ways to automate the home. Arduino (Uno, Mega, Pro, Mini) or Raspberry Pi can be used for automation. A simple example of controlling the light is explained in the following section. The components required are Arduino Uno, Relay Module, Bluetooth Module/Wi-Fi module, resistors (1 KΩ and 2.2. KΩ), and jumper wires.

Connect the components as given in the circuit diagram in Figure 4.10.

In order to upload the code, connect the Arduino through the USB port and open the Arduino IDE. Type the following program and upload the code to the Arduino Uno.

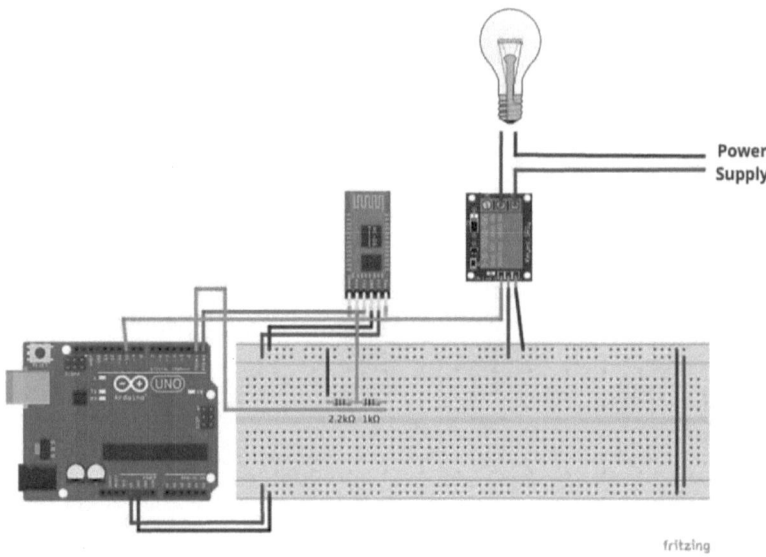

FIGURE 4.10   Circuit diagram – Automation of light.

```
#define RELAY_ON 0
#define RELAY_OFF 1
#define RELAY_1 4
void setup() {
      // Set pin as output.
      pinMode(RELAY_1, OUTPUT);
      // Initialize relay one as off so that on reset it
would be off by default
      digitalWrite(RELAY_1, RELAY_OFF);
      Serial.begin(9600);
      Serial.print("The Bulb will be on for 5 Seconds and Off
for 5 Seconds");
}
void loop() {
      Serial.print("\n"); //New line
      digitalWrite(RELAY_1, RELAY_ON);
      Serial.println("\n Bulb is now turned ON.");
      delay(5000);
      digitalWrite(RELAY_1, RELAY_OFF);
      Serial.println("\n Bulb is now turned OFF.");
      delay(5000);
}
```

After setting up the hardware and uploading the code, test whether the bulb switches on and off at interval of 5 seconds. If it is working, then the same code can be modified to read the input from the Bluetooth module and act according to the input received. For example, the code can be modified to switch on the bulb when the Arduino receives a value of 1 and switch it off when it receives a value of 0.

Any Android device can be configured to send commands to an Arduino using the Bluetooth module. For this, Arduino Bluetooth Controller app can be downloaded on any Android device. Install the app and give permission for the app to use the Bluetooth. After opening the app, select the HC-05 module, which is the Bluetooth module. Once the connection is successfully established with the HC-05 receiver module (Bluetooth module), the app will ask for the mode to be selected. Select the Switch Mode and in the settings give the value 1 in the ON text view and 0 in the OFF text view. Now the app will be ready to communicate the values to the Bluetooth receiver.

The above example is a simple example of using Arduino and a Bluetooth module to control the light. Additional appliances can be added by adding multichannel relay. The above example is given to aid in getting the idea of how home automation applications can be built. Proper safety precautions are to be taken while trying the above example as this might be dangerous because of the high voltage current that passes through the relay.

Any system proposed would have its own advantages and disadvantages.

4.3 ADVANTAGES OF IoT

- **Effective communication:**
    It is possible to improve communication over a network of interconnected systems, thus increasing the transparency of device communication and reducing inefficiencies.
- **Access information:**
    Data and information that are far away from our location can easily be viewed in real time. Due to network devices, an individual can access any information from anywhere in the world.
- **Accuracy:**
    The larger the data can be processed, the more accurate decisions can be made and activities can be completed based on those accurate decisions.
- **Automation:**
    Automation is a less time-consuming process without the need for human intervention, where parallel tasks can be performed at a time.

4.4 DISADVANTAGES OF IoT

4.4.1 COMPLEXITY

When the system is complex, there are chances of failure. You and your partner, for example, could receive messages that the milk is over and both of you could end up buying the same thing, which leaves you with double the quantity required. Or if the printer wants a single cartridge, there is a software bug that causes it to order the ink several times.

4.4.2 LESSER JOBS

The need for human labor will decrease considerably with tasks being automated, having a direct impact on employer life. As we move into the IoT future, the recruiting process of the professionals will decrease visibly.

4.4.3 COMPATIBILITY

No protocol has been developed to tag and track sensors as of now. A standard system such as the USB and Bluetooth is required, which is not so difficult to perform.

4.4.4 PRIVACY AND SECURITY

Each computer that a person uses is linked via the Internet in today's technological world. It increases the risk of any important data leakage. Sharing of information is a major drawback because confidential information may not be secure and can easily be hacked.

IoT Challenges in Security

- Unauthorized access of devices
- Software attacks
- Network breaches
- Encryption attacks

4.5 INDUSTRIAL IoT

IoT has different perceptions with the combination of other technologies and applications. In this section, we see a new perception of IoT. Industrial IoT (IIoT) is the application of IoT technology in industrial settings with respect to instrumentation and control of sensors and devices that engage cloud technologies. Currently, industries have started using machine-to machine communication (M2M) to accomplish wireless automation and control. But with the materialization of cloud and associated technologies, which may include analytics and machine learning, industries can pull off a new automation layer and create new income and business strategies. IIoT is commonly called to be the fourth wave of industrial revolution or Industry 4.0.

4.5.1 INDUSTRY 4.0

This is a phrase coined in Europe. It means the same as IIoT and refers to the fourth industrial revolution, as depicted in Figure 4.11. The term is interchangeable with IIoT and is now renowned internationally [17].

The widespread and well-known uses of Industry 4.0 are as follows:

1. **Preventive and predictive maintenance**: It is self-explanatory that prevention is better than sure. For instance, the heat of the machine, running time, and amount of fuel needed are all monitored in a smarter way.

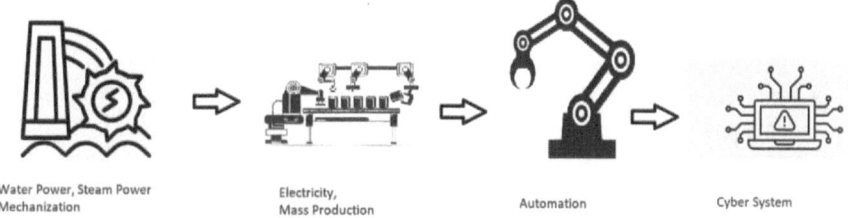

Water Power, Steam Power Mechanization

Electricity, Mass Production

Automation

Cyber System

FIGURE 4.11 Industry 4.0. (Courtesy: renaix.com.)

2. **Smart cities**: It is the biggest dream of many countries in this decade, and some have already started working toward that. It includes smart and healthy living of people, smart vehicles and traffic, and many more.
3. **Smart power grids**: Reduces the power usage with smarter machines.
4. **Smart manufacturing**: Reduces inappropriate and improper production.
5. **Smart and connected logistics**: Delivery of the manufactured goods are streamlined.
6. **Smart digital supply chains**: The entire delivery processes are monitored and ensured for quality.

In fact, IIoT influences the power of smart machines and real-time analytics to take advantage of the data that conventional machines have produced in industrial settings for these many years (Figure 4.12).

4.6 HOW IIoT WORKS

IIoT is a network of intelligent devices connected to form systems that monitor, collect, exchange, and analyze data. Each IIoT echosystem consists of:

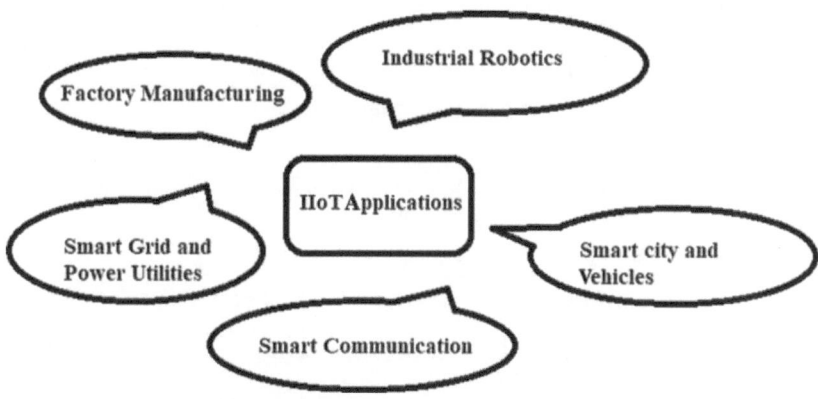

FIGURE 4.12 IIoT applications.

- Intelligent assets that can sense, communicate, and store information about themselves.
- Public and/or private data communications infrastructure.
- Analytics and applications that generate business information from raw data and people.

Edge devices and intelligent assets transmit information directly to the data communication infrastructure, where it is converted into actionable information on how a certain piece of machinery is operating, for instance. This information can then be used for predictive maintenance, as well as to optimize business processes. Figure 4.13 depicts the clear infrastructure of the IIoT.

4.7 BENEFITS OF IIoT

One of the top publicized benefits the IIoT affords businesses is predictive maintenance. This involves organizations using real-time data generated from IIoT systems to predict defects in machinery, for example, before they occur, enabling companies to take action to address those issues before a part fails or a machine goes down.

Another common benefit is improved field service. IIoT technologies help field service technicians identify potential issues in customer equipment before they become major issues, enabling techs to fix the problems before they inconvenience customers.

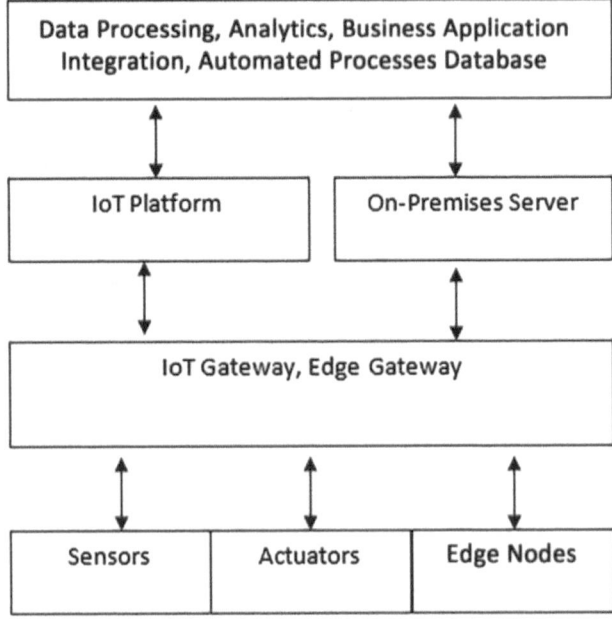

FIGURE 4.13   IIoT Infrastructure. (Courtesy: IoT Agenda-TechTarget).

Asset tracking is another IIoT perk. Suppliers, manufacturers, and customers can use asset management systems to track the location, status, and condition of products throughout the supply chain. The system will send instant alerts to stakeholders if the goods are damaged or are at risk of being damaged, giving them the chance to take immediate or preventive action to remedy the situation.

IIoT also permits enhanced customer satisfaction. When products are connected to IoT, the manufacturer can capture and analyze data about how customers use their products, enabling manufacturers and product designers to tailor future IoT devices and build more customer-centric product roadmaps.

IIoT also improves facility management. As manufacturing equipment is susceptible to wear and tear, as well as certain conditions within a factory, sensors can monitor vibrations, temperature, and other factors that might lead to operating conditions that are less than optimal.

4.8 IIoT VERSUS IoT

Although the IoT and the IIoT have many technologies in common, including cloud platforms, sensors, connectivity, machine-to-machine communications, and data analytics, they are used for different purposes.

IoT applications connect devices across multiple verticals, including agriculture, healthcare, enterprise, consumer and utilities, as well as governments and cities. IoT devices include smart appliances, fitness bands, and other applications that generally don't create emergency situations if something goes amiss.

IIoT applications, on the other hand, connect machines and devices in such industries as oil and gas, utilities, and manufacturing. System failures and downtime in IIoT deployments can result in high-risk situations or even life-threatening situations. IIoT applications are also more concerned with improving efficiency and improving health or safety versus the user-centric nature of IoT applications.

4.9 VENDORS IN IIoT

There are a number of vendors with IIoT platforms:

- Ability by ABB, a power and robotics company
- IoT System by Cisco, a networking company
- Field by Fanuc, a supplier of industrial automation equipment
- Predix by GE Digital, an energy management company
- Connected Performance Services by Honeywell, a software-industrial company
- Connyun by Kuka, a manufacturer of industrial robots
- ARM Ltd and Dassault systems
  (created in partnership with Infosys, an IT consulting firm)
- Wonderware by Schneider Electric, an energy management company
- MindSphere by Siemens, an industrial manufacturing company [18]

4.10   ENTERPRISE AND IIoT

An enterprise IoT platform is a type of IIoT software that allows organizations to securely manage all the connected people, systems, and things within an IIoT ecosystem. Both concepts IoT and IIoT have the same main character of availability: intelligent and connected devices. The only difference between those two is their general usages. While IoT is most commonly used for consumer usage, IIoT is used for industrial purposes such as manufacturing, supply chain monitor, and management system.

Enterprise IoT is the next advancement in technology that enables physical "things" with embedded computing devices to participate in business processes for reducing manual work and increasing overall business efficiency.

Taking advantage of a combination of technologies ranging from embedded devices with sensors and actuators to Internet-based communication and cloud platforms, the enterprise IoT applications can now automate business processes that depend on contextual information provided by programmed devices such as machines, vehicles, and other equipment. Further, such enterprise IoT applications can also send control instructions to these devices based on specific business rules.

Enterprise system and IoT will bridge diverse technologies to enable new business applications that connect with physical objects like devices and machines.

Enterprises have the capital, reach, resources, and the right reasons to deploy IoT solutions and services on a large scale. They will see benefits from the IoT increase fast enough to stimulate further adoption and investment.

The real value of enterprise IoT comes from data. With the advancements in Big Data processing and analysis, the insights derived from the IoT data sets will become priceless for the enterprise decision makers. Technologies such as real-time stream analysis and machine learning facilitate undeniable scenarios, including predictive maintenance and proactive monitoring of posh industrial equipment. Apart from mining the sensor data sets, connecting the legacy devices to software platforms and controlling them forms an essential element of an IoT solution.

Thus, industry and enterprise IoT are two precious components of IoT.

4.10.1   BLOCKCHAIN AND IoT

The recent catchphrase bitcoin and the related blockchain technology are making a high impact in the financial sector of the world but have not limited their applications to one sector. The blockchain is mainly associated with cryptography and the transactions involve data, hash, and the hash of previous block. The hashing function makes the data of a block to a fixed length content that can be understood as a unique fingerprint of the block [8]. Blockchain guarantees the safe storage and exchange of data in the common communication medium.

IoT has already started hiking our comfort and convenience by including smart objects in our life. Being smart may not provide increased security or privacy as the devices are always in the public network [9]. The security challenges and few other drawbacks of IoT technology can be surmounted or can be restrained with the block-chain technology with its enhanced features when both the technologies are used together. Some of the major parts of blockchain in IoT are data security, economy, transaction speed, decentralization, identity and access management, and resilient and reliable transaction.

Generally, with IoT, the smart devices are growing drastically. There should be better manageability and well-secured communication among the devices whether they are small devices such as sensor nodes to just sense and report or a smart control device that can make decisions to ensure integrity of the people involved. According to Gartner Investigation [10], the industrial approach toward IoT may involve few challenges. The idea of connecting billions of devices, controlling the decentralized devices, point-to-point communication among distributed devices, and securing the communication among devices in IoT are some of the challenges mentioned by Gartner. With the inclusion of blockchain in IoT, the issues of applying IoT technology can be mitigated considerably. Here we discuss some of the solutions of using blockchain in IoT.

4.10.1.1 Economy

The inclusion of blockchain in IoT makes the system more economical. IoT has centralized architecture with high-end servers, processors, and multiples of protocols, which incur higher overhead and maintenance costs. Since the blockchain is distributed in nature, the architecture cost is considerably low, which in turn benefits the IoT.

4.10.1.2 Transaction Speed

Blockchain has easy access of the blocks in a secured, reliable way, which can enable the automated exchange of data. IoT data access is speeded up with the blockchain because the third-party acceleration for the transactions is eliminated.

4.10.1.3 Decentralization

IoT, as discussed earlier, is a centralized technology that has its own drawbacks of overall cost, maintenance, and single point of failure. But the inclusion of blockchain in IoT makes it a decentralized system where the failures and errors in the transactions are not propagated. Figure 4.14 depicts the centralized (Figure 4.14a), decentralized (Figure 4.14b), and distributed (Figure 4.14c) network skeletons.

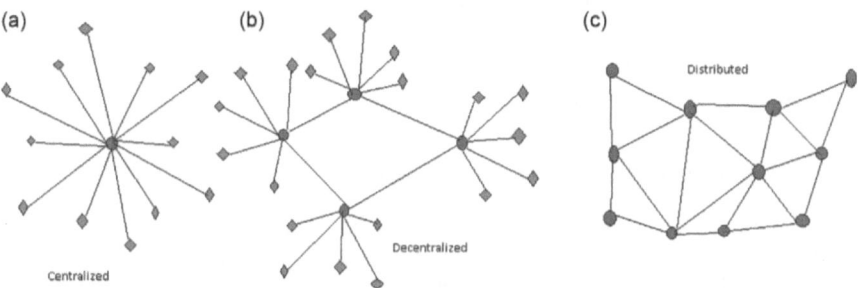

FIGURE 4.14 (a) Centralized, (b) decentralized, (c) distributed.

Generally, the centralized and decentralized systems depend on the control: single point of control or multiple point of control [11]. But the distributed systems deal with the location variation, where parts of the system are located in different physical locations.

4.10.1.4   Identity and Access Management

Identity and access management of blockchain improves the IoT security. IoT devices are commonly using private blockchains to store [7] cryptographic hashes of IoT devices state and configuration. This permanent record is used to validate the authenticity of the devices. Thus, the blockchain can provide defence against IP address forgery attacks and IP spoofing attacks as well as protects from fake signatures.

4.10.1.5   Resilient and Reliable Transaction

IoT devices in blockchain store distributed ledger copies on their memories that accumulate all transactions. Since each device holds the ledger, the sharing of information adds additional processing, storage, and power consumption to achieve better resiliency among the devices. Blockchain blocks are immutable; when included in IoT devices, they improve the traceability and accountability of the sensor devices and assure the proper message delivery.

4.10.1.6   Security

The blockchain is a distributed ledger in which each block is fingerprinted and time stamped to maintain its uniqueness. If one block content is changed, the fingerprint is changed and the block is no longer considered as the same block, and each block stores the hash of the previous block as well. It also ensures the security among the blocks by avoiding a changed block to be part of the chain.

4.10.2   COMPARISON OF BLOCKCHAIN AND IoT

The two major technologies that can be merged to promote their advantages in various applications have some dissimilarity. IoT is a centralized technology, whereas the blockchain is decentralized and distributed [12]. Resources are constrained in IoT but resource consumption is more in blockchain. Latency time is low in IoT but it is more in blockchain mining due to the distributed ledgers. IoT involves large number of devices, whereas the blockchain scalability is poor. IoT consumes limited bandwidth and resources as it contains the smaller devices, but blockchain consumes more bandwidth. The main difference that is always highlighted is the better security of blockchain technology than IoT.

4.10.3   APPLICATIONS OF BLOCKCHAIN IN IoT

Though IoT in its singularity has many applications, included blockchain technology is enhancing the power of IoT. Blockchain IoT is applicable for governments, smart monitoring, healthcare, commercial applications, agriculture, and communication systems. Here we highlight some of the applications of blockchain IoT.

1. **Healthcare:**

 The patient care and emergency support are enhancing with IoT, which can be secured with blockchain. The medical records can be consolidated with blockchain and saved securely in a way that cannot be tampered. Only authorized people can view the details for any history clarification, correct billing and medical claims from the corresponding insurance companies, and proper medication. This enhances IoT to be super smart in its performance.

2. **Supply chain:**

 The prevalent application in industry development of blockchain IoT is supply chain management and fleet monitoring. In supply chain management, all transactions are made transparent and traceable with the help of blockchain. It is common that the industries keep ledgers for their reference and records only; the outlets will not be monitored once they are delivered. But it is beneficial to keep track of the supply chain until the product has reached the end user to ensure the quality of the product.

 For instance, blockchain IoT can be used to track the complete cycle of a final product like potato, where it is cultivated, which variety of potato it is, which chips-producing company purchases it, in which part of the country it is sold, and the feedback of the consumers. The process has been shown in Figure 4.15. This kind of information is not only useful for the consumers to consume healthy and quality products, but it is also helpful for the manufacturers to improve the product in their own competitive way.

3. **Energy distribution:**

 Electricity is one of the inevitable parts of our lives. Nowadays the generation of electricity is not limited to government or private companies; even the home-generated electricity is also increasing with the solar power systems. So the utilization has to be streamlined with the consumers and producers. This can be done with blockchain IoT by maintaining a distributed ledger of the producer consumer details and making it transparent for the users. Energy distribution is not only for household and industry purposes but also for automobiles in the near future.

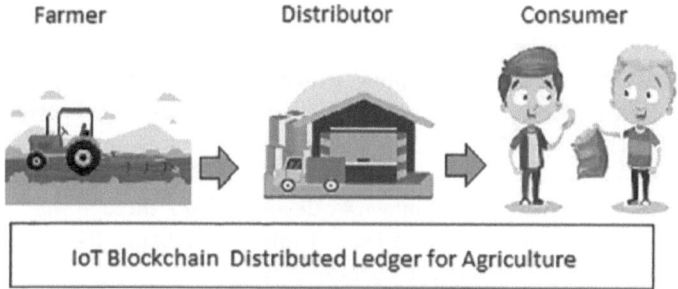

FIGURE 4.15 IoT blockchain-distributed ledger for agriculture for potato chips.

4. **Smart homes and smart parking solutions:**
   IoT devices are major parts of securing homes and making them smart. The insecure nature of few IoT devices and centralized features are overcome with the blockchain inclusion and security is highly enhanced. A company named NetObjex [13] has proposed a smart parking solution for automated payment with crypto wallets and for finding empty parking slots and ways in a highly crowded region with the application of blockchain IoT.

5. **Financial transactions:**
   The world is moving toward cashless, online transactions, which are in need of higher-end authenticity and authorization. Where technology grows, loopholes grow in parallel. Online transactions are well and good to avoid taking cash in hand and delivering to the receiver but the cybercrimes are also growing to threaten the growth of the technology. The blockchain IoT provides a better security and reduces the fear of online threats toward the financial transactions.

4.11   CHALLENGES OF BLOCKCHAIN IoT

Though the convergence of the two massive technologies has a lot of applications for the society, in reality, they have some challenges too. In this section, we discuss some of them. The noticeable challenges are scalability, energy consumption, processing time, and storage.

1. **Scalability:**
   The network is becoming bigger in size and spreading around the world, thus the transactions are also increasing. This is normal for an IoT, but when combining with blockchain, the limited bandwidth in the usage of cloud computing and processing of data in real time makes it cumbersome. The amount charged for the cloud services, including data storage, transaction fee are also increasing as the network is growing. This in turn has affected the growth of blockchain as it is in need of large storage and resources. Few companies are trying to overcome this defect by applying mini-blockchains, tree-chains, and side-chains [14]. Mini-blockchains are to reduce the scalability issues of blockchains that do not include high-bandwidth-occupying historic blocks.

2. **Energy consumption:**
   In the context of bitcoin mining, where the transaction records are added to the public ledger of the past transactions, the blockchain is always having high energy consumption. Generally, the IoT is with the devices that consume lesser energy and are battery-powered. Due to processing of more records, more energy is consumed by the blockchain, through which it does not match with IoT. With the help of mini-blockchains, the energy consumption can also be reduced considerably.

3. **Processing time:**
   Since the blockchain is more secured through cryptography, encryption, and hashing process, the processing time is becoming more. But in case of

IoT, it needs and holds lesser processing time. This difference in process time is a major flaw of the convergence.

4. **Storage:**

Blockchain-distributed technology doesn't use a central server to store the transaction details. It has distributed ledgers that each device involved has its own copy of the ledger; consequently, when the ledger size grows in size gradually, the storage needed by the ledger copies also grows. This growth of storage is also in par with the number of devices and transactions among them. Storage not only increases the cost of using it but also not coping up with minimal storage devices of IoT.

Some other legal issues may arise due to the crypto security and noncentral control of the authorization servers, as the IoT do not maintain such distributed security. Care must be taken to merge these two technologies to be pertinent for the real-time applications.

4.12 DISCUSSION ON APPROPRIATE USE CASES

Whenever new set of technologies are introduced, it is very challenging to determine the appropriate real-time scenarios in which the usage of those technologies make significant positive impacts. When the technologies are used just for the sake of utilizing an emerging trend, there would be plenty of bottlenecks and other practical problems that would adversely affect the business requirements. So here we have handpicked few use cases where the combination of IoT and blockchain are perfect partners in solving current business problems as well as providing assurance for future innovations.

4.12.1 SHARED ECONOMY APPLICATIONS

The combination of IoT and blockchain are most suitable for the shared economy applications where people and businesses come together in a decentralized environment such as Airbnb and exploit the opportunities of digital economy. In shared economy applications, people monetize what they have. Peer-to-peer automatic payment systems, management of digital rights, and exchange of foreign money or bonds can be few applications that are going to be influenced by the blockchain and IoT integration.

Peer-to-peer automatic payment systems and unconventional platforms for foreign exchanges are slowly getting attention due to the integration. We can see the rise of decentralized and autonomous applications – also known as Dapps. The existing buildings, household equipment, and other devices and vehicles can become part of Dapps by adding a sensor and network connection. By introducing blockchain, we can achieve transactions without a trusted third party by allowing our applications to function in a decentralized peer-to-peer mode. Already we have implementations in usage such as Slock, where a property or any unused asset can be linked to blockchain, and it can be monetized with the help of a Dapp (Figure 4.16).

A smart contract is a computer protocol intended to digitally facilitate, verify, or enforce the negotiation or performance of a contract. Smart contracts allow the

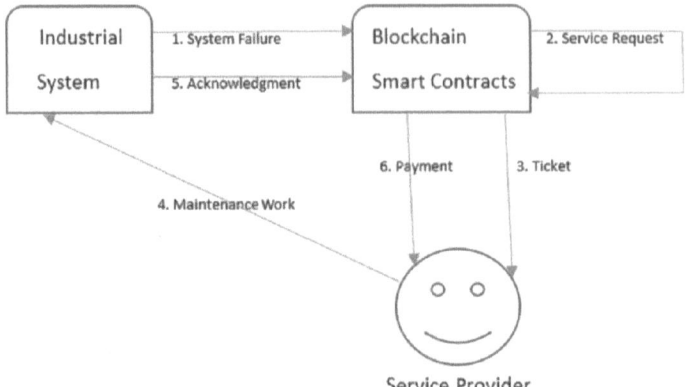

FIGURE 4.16   Smart contracts with IoT.

performance of credible transactions without third parties. These transactions are trackable and irreversible.

Proponents of smart contracts claim that many kinds of contractual clauses may be made partially or fully self-executing, self-enforcing, or both. The aim of smart contracts is to provide security that is superior to traditional contract law and to reduce other transaction costs associated with contracting. Various cryptocurrencies have implemented different types of smart contracts.

4.12.2   Supply Chain Management

Supply chain management is a complicated domain where several stakeholders participate from different places of their business. Collaboration between raw material suppliers, product manufacturers, logistics companies, shipments, retailers, and customers is crucial to increase the transparency of supply chain life cycle. To prevent losses and delays, the data collected from all the stages, starting from when the order is dispatched until the end-of-life of the product, is significant. Getting real-time visibility of the location and condition of shipments is a primary factor that decides the effectiveness of supply chains and the revenue of the companies. However, it has been immensely challenging to get all these data on time to make the business more profitable and to reduce the losses.

The contracts and rules exchanged among the different parties are the essential components of supply chain services. They are signed before the operation begins. Once the shipping is started, parties such as shipping agencies, ports, custom officials, carriers, and other entities take part in the execution. At each level, set of acknowledgements, messages, and receipts are exchanged. This long chain of data is well suited for making use of blockchain to register all the transactions in between to ensure reliable, data-driven smooth process. The distributed ledger will be verified by the customers, logistics providers, and government authorities anytime to obtain reliable information and to have better control over the entire supply chain and logistics domain.

4.12.3 SMART CITY APPLICATIONS

Smart city is a domain where plenty of smart solutions, including smart infrastructures, smart buildings, are part of. In smart buildings, the information from sensors and control information from authorized users are sent to a centralized server and directed for appropriate device. In smart houses, the household devices such as refrigerators, dish washers, lights, and security systems such as alarms, cameras, and locks are present. All these devices can be connected to the Internet and controlled by Internet. Ensuring reliability and preventing intrusion are biggest challenges in this problem space. Because we allow the authorized users to monitor and control from remote sites using handheld devices or web applications, blockchain can ensure the privacy and security of these applications. The communication between devices and control messages can be recorded in the distributed ledger.

The information can be verified at any point in time as there are no chances of corruption or deletion in the distributed ledgers. Any unauthorized transactions will be rejected automatically. Algorithms such as AES would provide symmetric cryptography for confidentiality, and hash functions can be used to verify the integrity of communication. All these would make the smart infrastructures robust (Table 4.1).

The next section discusses a simple architecture that explains about the converged blockchain IoT.

4.12.4 BLOCKCHAIN IOT ARCHITECTURE

Figure 4.15 explains how IoT and blockchain, two different technologies, can be converged to employ for different applications. IoT refers to a loosely coupled system of multiple heterogeneous and homogeneous devices that can sense, process, and network. As explained earlier, a fundamental problem with current IoT systems is their architecture, with a centralized client-server model managed by a central authority. This makes it susceptible to a single point of failure. Blockchain addresses this problem by decentralizing decision making to a consensus-based shared network of devices.

The blockchain can be deployed in three domains:

1. **Public**: bitcoin and Ethereum come under this category. Each and every node can send or read transactions without requiring any permission. Consensus is open to the public.
2. **Consortium area**: It comes under partial permission. The permission to read or send may be made public or may be provided only to few authorized nodes.
3. **Private**: Only the organization to whom the network of blockchain belongs can write transaction to it.

The privacy of transaction history in the shared ledger for a network of IoT devices cannot be easily granted on public blockchains, because transaction pattern analysis can be applied to make inferences about the identities of users or devices behind

TABLE 4.1

Appropriateness of Blockchain [15]

Scenario	Go for Blockchain	Go for Other Options
Application is data-intensive and it is used by huge number of users	If all the users have write access to the database and there are problems of conflicts	If only a limited number of users have write access, you may use a centralized database
Applications where multiple users involve in transactions	If a mutual trust is required among the users, blockchain would make the validation of transactions simple	If mutual trust is not required, distributed database can be used instead of blockchain
Applications using third-party authorities	If the involvement of third-party authorities is creating bottlenecks, blockchain is recommended	If third-party entities are making the process simple without overheads, they would be preferred
Applications with stable requirements	If the application requirements would not undergo major changes in the future	If there are very frequent updates in the core application, blockchain may become an overhead
Availability of skill sets and other resources	When the company is fully equipped with required skill sets and potential to pursue with advanced technologies can go ahead with blockchain	Companies that are unable to invest on multiple skill sets and running on strict budget may not find blockchain feasible
Relationship with stakeholders	If the internal and external stakeholders are ready to adopt technical and organizational challenges of blockchain integration, blockchain can be considered	If the stakeholders may not be able to quickly adopt to blockchain or if they have threats for continuing in future, it is not advisable to go for blockchains

public keys. Organizations should investigate their privacy requirements to see whether hybrid or private blockchains might suit their requirements better.

4.12.5 Network Layer

The IoT connected directly with network layer. The main objective of network layer is to handle operating system and message passing protocols. Nodes/devices are linked with the networks.

4.12.6 Distributed Ledger Technologies (DLT)

A distributed ledger (also called a shared ledger or distributed ledger technology or DLT) is a consensus of replicated, shared, and synchronized digital data geographically spread across multiple sites, countries, or institutions [3]. There is no central administrator or centralized data storage. One DLT is bitcoin, which is a cryptocurrency. It is a decentralized digital currency without a central bank or single administrator that can be sent from user to user on the peer-to-peer bitcoin network without the need for intermediaries. Transactions are verified by network nodes through cryptography and recorded in a public-distributed ledger.

Litecoin uses an open source software to create and transfer coins, and is decentralized.

The MultiChain technology is a platform that helps users to establish a certain private blockchain that can be used by the organizations for financial transactions. A simple API and a command-line interface are what MultiChain provides us. This helps to preserve and set up the chain.

Corda is an open-source blockchain platform that enables businesses to transact directly and in strict privacy using smart contracts, reducing transaction and record-keeping costs and streamlining business operations.

Ethereum is an open-source public blockchain-based distributed computing platform and operating system featuring smart contract (scripting) functionality. Ether is a cryptocurrency generated by the Ethereum platform, and it is used to compensate mining nodes for computations performed.

Ripple is built upon a distributed open-source protocol, and it supports tokens representing fiat currency, cryptocurrency, commodities (Figure 4.17).

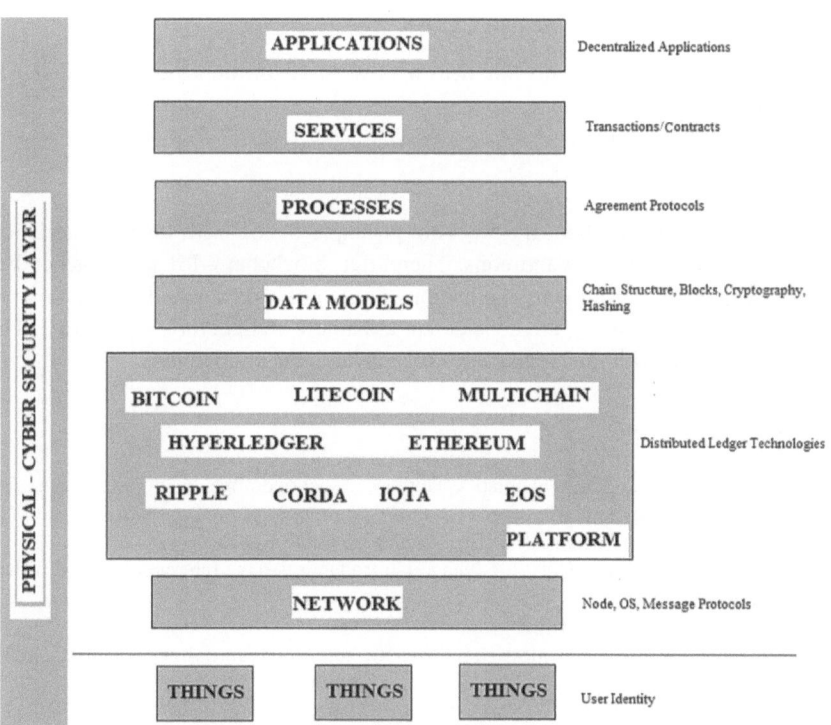

FIGURE 4.17   Blockchain–IoT convergence architecture.

4.12.7   DATA MODELS

Data models in blockchain technology are the current state of the ledger. The retrieved data are hashed and stored in the form of blocks, which are decentralized and cryptographically secured.

4.12.8   PROCESSES

This layer deals with the consensus protocols, which provide the proof of work (PoW) and verify the action in the networks.

4.12.9   SERVICES

The contracts and transactions between peers are dealt in this layer. Contract is used to verify and validate the participants of the network.

4.12.10   A DECENTRALIZED APPLICATION (DAPP, dAPP, DApp, OR dapp)

It is a computer application that runs on a distributed computing system. Dapps have been popularized by DLTs.

4.12.11   PHYSICAL-CYBER SECURITY LAYER

It comprises of the components that deal with security for the entire blockchain IoT architecture.

4.13   SUMMARY

This chapter discussed about the working principle of IoT with its characteristics, applications, and their pros and cons. Thereafter, blockchain–IoT convergence has been discussed with prominent applications of the convergence and the challenges. Some of the design issues were considered and a new architecture is proposed. The chapter concluded with the blockchain–IoT convergence architecture.

REFERENCES

1. Malathi Balaji, Shafique Ahmad Chaudhry, "A cooperative trilateration technique for object localization" *20th International Conference on Advanced Communication Technology (ICACT)*, Feb 2018, IEEE Publisher, South Korea.
2. www.codeproject.com, Oct 2014, "Stage 1-Introduction to the Internet of Things: What, Why and How".
3. https://en.wikipedia.org/wiki/
4. https://newcities.org/cityquest-songdo-south-korea-conceptualized-ultimate-smart-sustainable-city/
5. www.tutorialspoint.com
6. https://www.linkedin.com/pulse/internet-things-iot-characteristics-kavyashree-g-c
7. http://www.techgeekbuzz.com/iot-applications/

8. Mohammad Maroufi, Reza Abdolee, and Behzad Mozaffari Tazekand, "On the convergence of blockchain and internet of things (IoT) technologies", arXiv:1904.01936v1 cs.DC., 11 Mar 2019.

9. David Nettikadan, Riya T Raphael, Blessy Daise Paul, "Converging blockchain and internet of things", *International Journal of Innovative Technology and Exploring Engineering (IJITEE)* May 2019, 8 (7S2), 2278–3075.

10. Leif-Olof Wallin, Mike Walker, Nick Jones, "How to address the top five IoT challenges with enterprise architecture," Gartner.Inc, Online, November 2016.

11. https://medium.com/nakamo-to/whats-the-difference-between-decentralized-and-distributed-1b8de5e7f5a4

12. Hany F. Atlam, et.al, "Blockchain with internet of things: Benefits, challenges, and future directions", *International Journal of Intelligent Systems and Applications*, 2018, 6, 40–48.

13. https://hackernoon.com/blockchain-and-iot-bringing-transformation-to-the-world-2f69cb0c498a

14. https://coinrivet.com/what-is-the-mini-blockchain-scheme/

15. G. Selvakumar, S. Hemalatha, "A study on integrating IoT Applications with Blockchain" *2019 International Conference on Computer Communication and Informatics (ICCCI-2019) IEEE Xplore*, Jan 23–25, Coimbatore, India, 2019.

16. R. Dharshini, A. Mohana Priya, S. P. Kavya, G. Divya, S. Bharath, S. Guna Sekaran, "Garden monitoring system and controlling pesticide using IoT", *International Journal of Research in Advent Technology*, 2019, 7 (3) 2321–9637.

17. https://wikipedia.org

18. https://www.openpr.com>news>industrialIoT

5 Research Issues and Solution in Blockchain Healthcare

R. L. Priya and S. Vinila Jinny
NICHE

CONTENTS

5.1   INTRODUCTION

Blockchain technology (BT) is mainly designed to solve various problems in almost every sector. Blockchain can be defined as a decentralized incorruptible digital ledger technology. This emerging technology has a promising potential to transmit securely any digital form of data or coins between stakeholders. It is basically used when there is a presence of multiple stakeholders and to ensure the transparency and integrity of data in the network. The information related to every transaction of such distributed network is shared and made available to all nodes in a secured manner. The advantage of such an emerging technology is easy retrieval of trusted information through distributed network.

In the era of Information and Communication Technology (ICT) and Internet of Things (IoT), their usage in the health sector has become a promising challenge for researchers. Many researchers have already started the exploration of BT toward healthcare domain. Nowadays, the major challenges faced in healthcare domain are security, interoperability, privacy, sharing, authentication, and policy aspects. Hence, it is essential to define an efficient, unique, sustainable, and standardized information business model for the same.

5.2   BACKGROUND

Here, we discuss about the various research challenges and their current state of research in healthcare domain related to BT. There are numerous business models, and medical applications are defined in healthcare industry, which varies from country to country. There are some specific healthcare applications that have massive opportunities and benefits of replacing the existing IT infrastructure with BT. Dave et al. define three different ways through which BT brings transformation into healthcare technology. The first option is to build blockchain-based electro records to generate a patient's database, which will be accessible to every department of healthcare [1]. The second option is to find a novel solution in medical supply chain management using blockchain to ensure smooth and safe transition of products. The third option is to deploy blockchain concepts in the genomic market for secure sharing of billions of data. The main advantage of integrating blockchain in healthcare is to solve inefficiencies that exist in the current system. Moreover, blockchains give a promising change to healthcare, which acts as a key catalyst for healthier communities.

5.3 RESEARCH STATE OF BLOCKCHAIN IN HEALTHCARE

The adoption of blockchain in healthcare and other biomedical applications is to improve authenticity, transparency, availability, robustness, auditability, security, privacy, integrity, and encourage decentralized management. Decentralization is a very important aspect in blockchain as it provides a way to create a democratic decision process. Additionally, it gives confidence to the term "For the People and By the People". Many healthcare organizations and research academia are working on the use of blockchain in medical network. We shall next discuss about some of the popular applications of ledger technology and how they are applied in healthcare domain. These applications provide efficient solutions to the research challenges for the following categories.

5.3.1 CLINICAL DATA MANAGEMENT/PATIENTS CARE MANAGEMENT

In healthcare management, it is very essential to streamline patients' care by quicker diagnoses and provide personalized care plan. This in turn will reduce the time and cost of access of patients' health records by doctors or other health professionals. Some applications are enfolding the concept of blockchain clinical records for making personalized care plans and creating secured and sharable patient health database.

5.3.1.1 SimplyVital Health (Watertown, Massachusetts)

SimplyVital Health is designed by Nexus Health Platform [2] as an open-source database. It allows authenticated healthcare professionals to access patients' pertinent information that employs blockchain security in patients' clinical database. Thus, it helps doctors and clinicians to access medical information of patients even in emergency situations more quickly than any traditional methods.

5.3.1.2 Hashed Health (Nashville)

Hashed Health [3,4] is a blockchain adopted for healthcare. It provides distributed ledger technology (DLT) solutions for many business problems. Some of its product innovations in healthcare are Signal Stream, Professional Credentials Exchange, and Bramble.

5.3.1.3 Coral Health (Vancouver, Canada)

Coral Health [2] helps to accelerate the patient care process using BT. It builds smart contracts between healthcare providers and patients to ensure the availability of data. It promotes automatic administrative processes in order to attain accurate treatment and improve health outcomes.

5.3.1.4 Robomed Network (Moscow, Russia)

Robomed [2] uses artificial intelligence (AI) on the blockchain to build a decentralized network for better patient care management. It supports various tools such as wearable devices, telemonitoring, and chatbots for collecting patients' information in a secured manner. This medical information is securely stored and shared with the patient's medical team.

5.3.1.5 Patientory (Atlanta, Georgia)

Patientory [2] uses end-to-end cryptography technique via blockchain to transfer patients' clinical information in a secured way. It ensures secure storage and sharing of important medical information.

Similarly, many more projects such as MedVault, Healthcare Data Gateways, Fatcom, BitHealth, and so on focus on exchanging patient care data among health providers using blockchain ledger technology in order to improve patients' clinical record management.

5.3.2 PHARMACY SUPPLY CHAIN MANAGEMENT

The supply of pharmacy products and traceability of drugs is one of the most prevalent use-case, required transformations in blockchain. In the healthcare industry, clinical supply products, medications, blood products, and health devices are few examples where blockchain ledger technology is created. It helps to record each and every transaction of data in the supply chain network. As it is a distributed network, it guarantees for better transparency in the process of health supply chain management. Next we discuss some applications of health supply chain management deployed in blockchain.

5.3.2.1 Chronicled (San Francisco, California)

Chronicled network [2] created the Mediledger system in the year 2017 to ensure privacy, safety, and efficiency of medical supply chain management. The blockchain-based network helps pharmaceutical companies get medicines in a safe and efficient way. It can track drug trafficking, law enforcement for reviewing suspicious activities during drug shipments.

5.3.2.2 BlockPharma (Paris, France)

BlockPharma uses blockchain-based supply chain management system. The mobile application was developed to make people aware and alert them (especially patients) when taking falsified medicines. The application [2] works by scanning the supply chain and performing verification in order to prevent patients from taking fake medicines. This technology helps to remove almost 15% of fake medicines in the world.

5.3.2.3 Tierion (Mountain View, California)

Tierion makes use of BT to maintain a clear possession history of medicines by conducting formal reviews. Formal review performs auditing of documents, records, and medicines. It helps to maintain proof of ownership using time stamp and credentials within the pharmacy supply chain network. Recently, Tierion proposed a new application, "Multi-network Coin" [2], which helps to make bitcoin more applicable and adaptable.

5.3.2.4 Centers for Disease Control and Prevention (CDC) (Atlanta, Georgia)

CDC is designed to monitor diseases using BT. This application uses blockchain's valuable parameters such as time stamp, peer-to-peer patient health status report, and processing capabilities to yield better real-time report. Recently, in collaboration with IBM, it is working on blockchain-based surveillance system for effective gathering of patients' data.

5.3.3 BREAKTHROUGHS IN GENOMIC MARKET

In the current scenario, according to laws of several countries, no organization has a transparent legal owner for genomic knowledge. The problems faced in existing genomic market are obtaining proper genomic information, privacy, and high costs. Now, the world discusses the solution for genomics by blockchain-based genetic analysis. There is a great contribution of Medical Genomics (Boston) [5] toward using blockchain for proof of existence. One of its major focuses is on genomic process, where analysis helps to extract information of patients with seizure disorders based on similar conditions and endocannabinoid system. Similarly, there are some more projects working on genomics, which are discussed next.

5.3.3.1 Nebula Genomics (Boston, Massachusetts)

Nebula [2] is a perfect model working to build an encrypted form of wider genetic database providing access to valuable information for the essential stakeholders. Such decentralized ledger technology helps to remove the concept of expensive middlemen in genomic network. Hence, it encourages users to safely sell their genetic data.

5.3.3.2 Encrypgen Gene-chain (Coral Springs, Florida)

Encrypgen [2] network uses blockchain-backed platform to facilitate certain functionalities in genomics such as searching, sharing, storing, buying, and selling of genetic information. The chain network protects its users' privacy by creating safe and traceable DNA tokens. Currently, Gene-chain network is working toward expansion of building user profile with the inclusion of self-reported clinical and behavioral data.

5.3.3.3 DOC.AI (Palo Alto, California)

DOC.AI [2] applies AI to define decentralized medical data network on the blockchain. It does not save any patient data in the chain model, rather once the information is used, data are eliminated completely to ensure privacy and security. Currently, it is working with health insurer Anthem for building the prediction model of the occurrence of allergic reactions.

5.3.4 IMPACT OF BLOCKCHAIN IN HEALTHCARE

BT can be applied in the healthcare sector because of its data handling and decentralization features. Moreover, it deals with authenticity, confidentiality, security, and

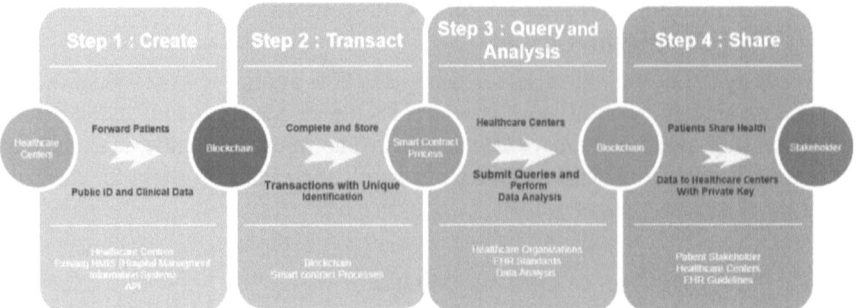

FIGURE 5.1   Impact of blockchain in healthcare sector.

privacy of patients' sensitive data with minimal cost and time. The healthcare data include patients' compiling episodes, lab reports, medical history records, medical analysis, and others. The major functions of BT in the healthcare domain are to provide secure transmission and efficient storage of patients' data to various stakeholders such as patients, doctors, and health professionals [4].

Figure 5.1 depicts various impacts of decentralized ledger technology in healthcare domain. These are divided into four different sections. The first section describes the ability of blockchain platform to create secured relationship between patients and healthcare providers for smooth transmission of medical data or information. Different healthcare organizations provide efficient services to patients and manage the clinical data using ledger technology. Second, blockchain has the capability to handle every transaction using stored patients' identity. To support trusted transactions, blockchain allows sharing of smart contract among anonymous parties.

Third, it gives healthcare organizations secure access and quicker retrieval of patients' medical data. The available medical data are used to extract useful medical information by evolving new patterns. Finally, such a DLT uses cryptography method to share sensitive patient information with various healthcare providers.

There are further uses of blockchain applications developed by popular companies like Microsoft, IBM, Deloitte, and Accenture [4] that work toward security and other features of the healthcare sector.

5.3.4.1   BURSTIQ (Colorado Springs, Colorado)

BurstIQ application [2] is developed for healthcare providers using blockchain to ensure safe and secure transmission of huge amounts of patient data. It manages and maintains the complete up-to-date health information of patients as per Health Insurance Portability and Accountability Act (HIPAA) compliance.

5.3.4.2   Factom (Austin, Texas)

Factom [2] employs blockchain platform to store patients' digital records in a secured factom chips. Such applications are used by the healthcare industry to hold private patient medical information, and these digital records are accessible only by certain authorities such as hospitals and healthcare administrators of healthcare companies.

5.3.4.3 Medical Chain (London, England)

Medical Chain [2] platform maintains the history of health records and provides protection to patients' identity from outside sources. It released another application, known as Myclinic.com, in May 2018. MyClinic.com creates a patient–doctor platform for video consultation. The pay module is implemented in BT with "Med Tokens".

5.3.4.4 Guard Time (Irvine, California)

Guard Time [4] uses the blockchain platform for cybersecurity applications. It has collaborated with Verizon Enterprise Solutions to deploy various services based on Guard Time's Keyless Signature Infrastructure (KSI) blockchain.

5.3.4.5 MedRec (MIT Media Lab)

MedRec [6] helps to build a patient–provider relationship using smart contract based on DLT. Smart contract is the agreement term made between the involving parties, written in the form of code. It is used as a decentralized medical record management system and provides access to patients' medical history. Initially, MedRec was implemented using Ethereum blockchain. The major benefit of using such a private blockchain is to track only the genesis not an entire Ethereum.

5.4 STAKEHOLDERS OF HEALTHCARE

In the current scenario without blockchain platform, many researchers, healthcare providers, and payers are facing difficulty in accessing medical data. Moreover, being sensitive information, most of the healthcare providers do not share clinical notes or other forms of patient medical data. The DLT in healthcare guarantees security and privacy of healthcare data. Based on the survey report in Ref. [7], nearly 189,945,874 healthcare records were either lost or stolen during 2009–2018.

Stakeholders play an eminent role in healthcare sector. The Figure 5.2 describes the major stakeholders involved in the healthcare domain. They are patients (in patient and out patient), healthcare providers (hospitals, healthcare centers, doctors, clinicians, lab technician, etc.), data analysts, and claim or insurance providers. For detailed analysis and proper diagnosis of health records, coordination and communication among these stakeholders are very essential. This information is later used by researchers and AI experts to improve the performance of the healthcare system.

5.5 STATE-OF-THE-ART FOR HEALTH RECORDS

There are diverse and tremendous amounts of medical records available from different sources derived either by traditional methods, collecting patient data in text, medical images or X-ray reports and ultrasounds or by latest smart technologies, collecting real-time patient data using Iota-based healthcare devices, trackers, AI, and so on. In the healthcare sector, adoption of universal standards in Electronic Health Records (EHR) is of great challenge. In the United States, EHRs are transferred as per standards and requirements specified by HIPAA security and privacy rules.

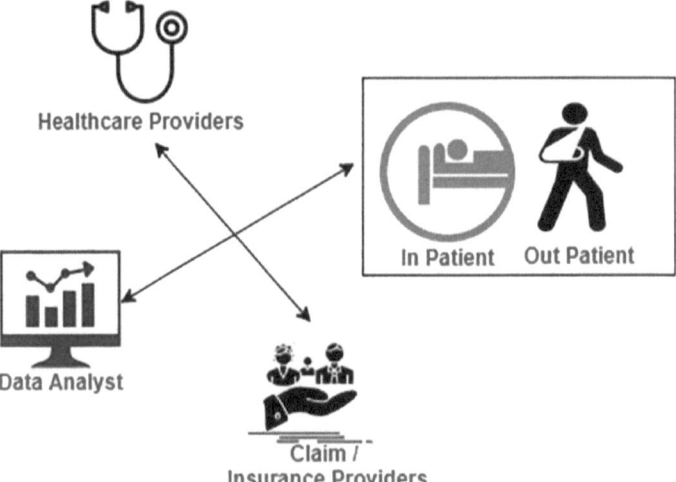

FIGURE 5.2   Stakeholders of healthcare system.

EHR is a system of collecting patients' health-related data in digital form. Similarly, Personal Health Record (PHR) helps in collecting patients' digital clinical records. As defined by Lee at al. [8], EHRs are of two types: Health data collected by smart devices or sensors and Electronic Medical Records (EMRs).

5.5.1 ELECTRONIC MEDICAL RECORDS

EMR is a collection of healthcare-related information gathered by different medical institutions. Such information is acquired as a series of time-specific information of a patient. The list of EMR data recorded by hospitals and other healthcare centers is shown in Figure 5.3.

Tanwar et al. [9] suggested a novel approach to embed blockchain network for EHR record sharing. Such systems are defined with four stakeholders: patient, clinician, lab technician, and system administrator. The proposed system uses smart contracts, and the distribution of EHR data to other stakeholders are achieved by shared symmetric key and private key in block transaction. Various algorithms are proposed for access control management of four different participants. The distribution of EHR data is done using hyperledger fabric and hyperledger composer. The survey in Ref. [10] discusses that the wearable devices are best suited for monitoring purposes.

Another extended version of e-health and m-health applications is known as S-Health (Smart Health application). It is designed to retrieve medical data from EHR, PHR, and smart cities infrastructure in order to develop a relevant feedback system. Such types of systems should be built on trust; it is essential to address certain security issues. To solve certain security issues of s-health apps, a blockchain-based framework is suggested in Ref. [10] to guarantee trust using IoT and 5G.

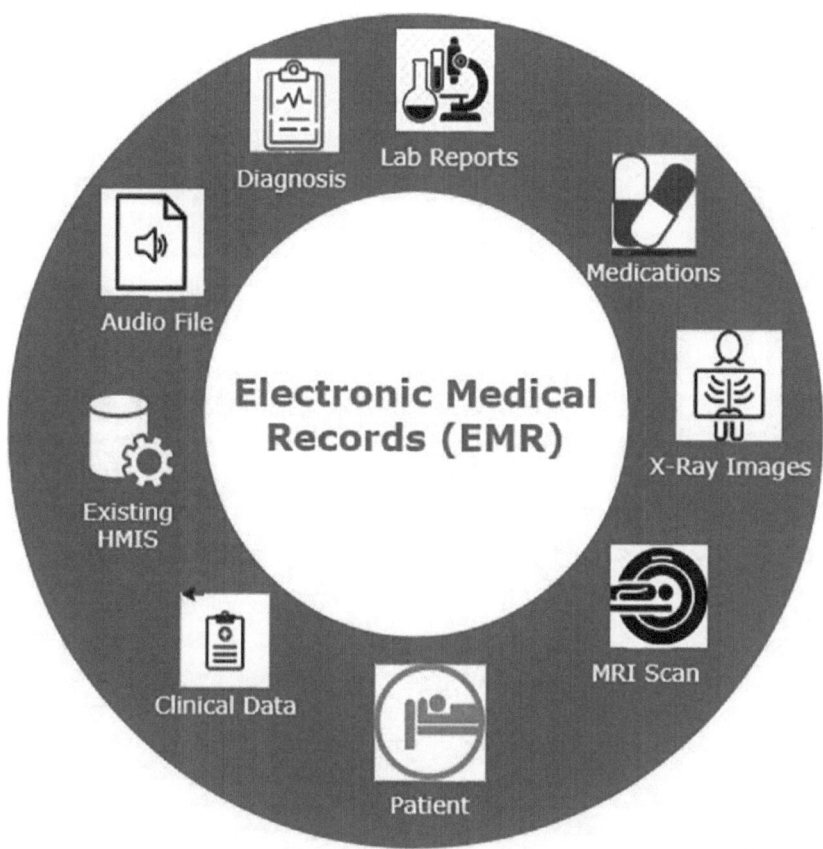

FIGURE 5.3 Electronic medical record.

5.5.2 IoT Sensors or IoT-Based Healthcare Devices

Today, various smart devices are available in the form of human wearable gadgets or as smart watches in healthcare industry. These devices are designed especially for remote patient monitoring and to monitor various health parameters of the patient. In healthcare, IoT has been identified as subsector of Internet of Healthcare Things (IoHT) [11]. Various sensors used to retrieve medical data in IoT healthcare field are listed in Figure 5.4.

Embedding of BT into IoHT helps to maintain the privacy of medical records. Intelligent sensor nodes [12] can accurately monitor patients' primary health parameters such as temperature, respiratory rate, heart rate, blood pressure, and so on. Such IoT-enabled devices can be easily deployed anywhere to obtain a patient's detailed information. This smart device helps to track the behavior change of patients and gives alert notification at the time of emergency to healthcare providers.

Badr et al. [13] propose a new protocol known as Pseudonym-Based Encryption with Different Authorities (PBE-DA) to guarantee the validity of IoT–EHR system

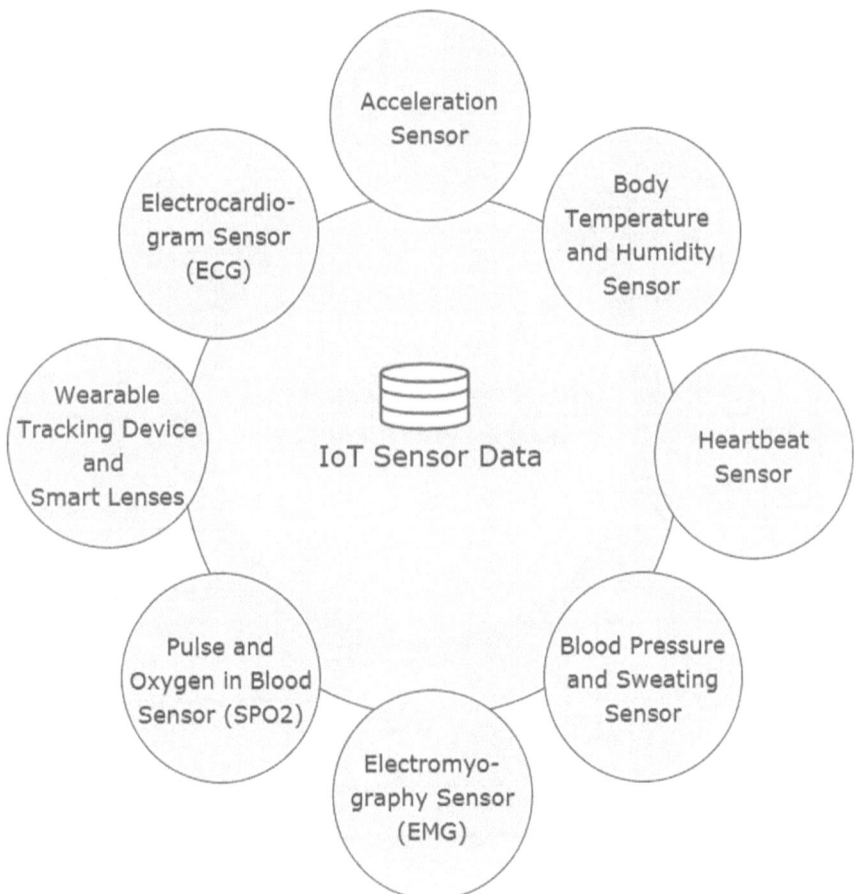

FIGURE 5.4   Different kinds of IoT health data.

using blockchain-based multitier framework. As per the previous study, Ref. [14] discusses consortium-based blockchain system for the addition of new healthcare data of a particular patient in an efficient way. Such a system exploits immutability using blockchain and IoT in healthcare data.

5.6   USE CASES AND ITS SOLUTIONS IN HEALTHCARE USING BLOCKCHAIN

The revolution of blockchain in the healthcare domain has transformed the world of technology into a new distributed and secure infrastructure. Various characteristics and features of blockchain [15–18] encourage domain experts in solving many problems, where a middleman is replaced with this recent BT. With the existing healthcare infrastructure, many researchers and health practitioners today are struggling with communication delay, reluctance in sharing medical information, and trustless transaction. Moreover, incompatibility of vendor-specific healthcare system creates a big gap in the healthcare communication world. Hence, it is very difficult to define secured and trusted communication among stakeholders and to provide efficient patient-centric care.

The benefit of defining use case scenario in healthcare field using BTs/DLT [19] is to ensure availability and accessibility of any medical data (including software and hardware) and to improve tracking and disease management.

In this digital era, DLT plays an eminent role in authentication and verification processes of the healthcare image sharing and security features. In addition, BTs are widely used in other application areas of healthcare, such as radiology [20], for tracking medical education, certification, and credential verification of health workers.

5.6.1 Use Case for Healthcare Providers and Insurance Companies

Health insurance is a type of insurance that covers the medical expenses of a person. It may cover the cost of treatment provided at medical emergency, long-term treatment of any chronic disease, and minor or major accidents or surgical cost, based on insurance coverage [21].

In any health organization, it becomes a general practice to share original bills between healthcare providers and insurance companies on the approved budget and schedule. During claiming process in any business sector, there are more chances of data manipulation, delay in closing bill settlement, and fraud transactions by some higher authorities or middle man (third party/agencies). BT is largely pioneered by the concept of decentralization like bitcoin, which impacts a lot in secured financial transactions. In order to avoid bill transaction problems, operate the application in a fully decentralized manner and trace the history of transactions between the authorized parties [22].

Saeedi et al. [18] describe a novel method – Claim chain application to avoid the fraud transactions in a bill-claiming process using blockchain. Claim chain app is developed using scrum process model and satisfies the requirements of current healthcare practice in Saudi Arabia. The software development life cycle (SDLC) of scrum framework is used to identify the weaknesses, and it brought the adoption of blockchain into the framework. The current practice of claiming process is depicted using Business Process Model and Notation (BPMN) [18].

In the conventional process, after the patient presents his or her insurance identity card at the hospital reception desk, policy details are verified and validated. Based on the approval of the policy, treatment bill is delivered to the stakeholder. Generally, hospital bills are delivered to the stakeholders on monthly basis. Thereafter, the bills are verified and payment is released on an insurance claim [18].

The abovementioned traditional method has now been replaced by blockchain ledger technology to ensure secure transfer of treatment bills from healthcare centers or hospitals to insurance company as shown in Figure 5.5. Claim chain app [18] is designed mainly using three major components: bill generator, blockchain, and bill retriever.

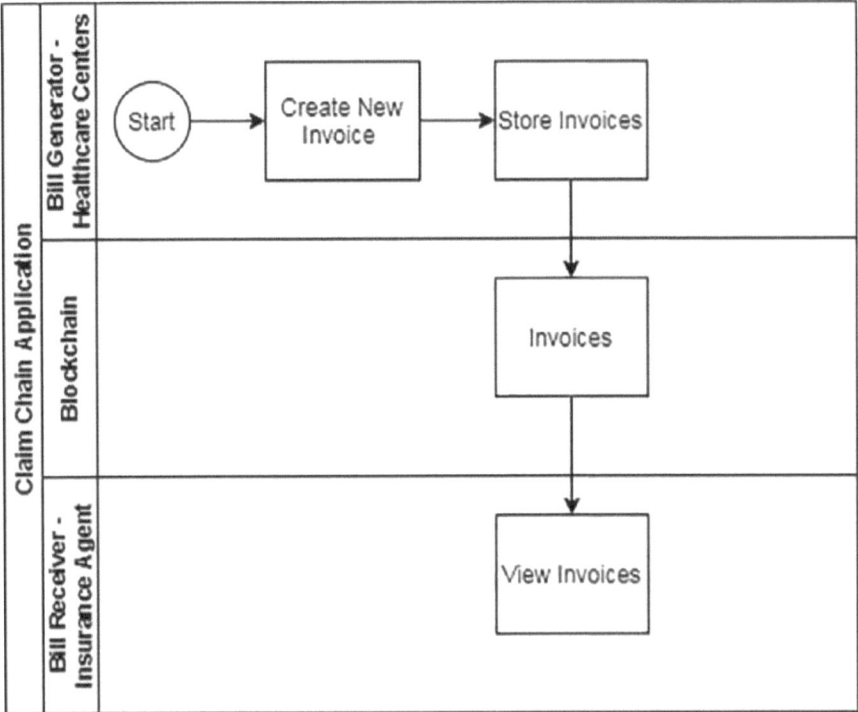

FIGURE 5.5   Claim chain process using blockchain.

Most of the claiming procedures are as same as the traditional process. Once the bill is generated by the healthcare center, it is then passed to the insurance provider using blockchain. The blockchain is built using several blocks consisting of bill information and connected in chains by mining process. Here, the claim details (patient identity certificate, services, bills, etc.) are passed only to authorized insurance companies by applying encryption and hashing techniques. Hence, the manual revision process of the claim is reduced a lot.

5.6.2   Use Case for Opioid Prescription Tracking

Opioid is a type of medication used for pain relief at the time of anesthesia, post-surgery recovery, physical disorders, etc. Today, many healthcare professionals and pharmacies are encouraged to prescribe opioid [21]. But the pain relief is only for a short period and, moreover, it adds to the addiction cost. Furthermore, health providers are more inclined toward costs incurred in the treatment and return profits. Similarly, pharmacies aim to sell and distribute more opioids in the market for greater return profits to shareholders.

BT is used to establish a trusted and secure network model for storing and tracking opioid-specific transactions between healthcare centers and pharmacies [21]. Such a distributed technology can be developed on a shared and permission blockchain platform. The consortium welcomes new members to the system based on mutual agreement of predefined rules. This model provides the remedy for the existing problem by obtaining complete opioid prescription history to detect prescriptions and data accessibility to improvise patient care.

5.6.3   Use Case for Telemedicine and Patient Care

Telemedicine is also referred as "telehealth", a new trend in technology for providing various medical services to those patients who live in remote areas and use telecommunication. This technology is used by patients living in remote areas where limited medical facilities and doctor or clinical expert consultation is available; telemedicine in such cases provides doctor or clinical expert consultation much faster and in an efficient way, without much travel. Such a system is more beneficial for most of the developing and developed countries in terms of cost reduction and better patient-centric care.

Nowadays, numerous smart mobile applications and telemedicine devices are available with 24/7 accessibility offered either by government or by private agencies for continuous care to patients. These user-friendly apps and devices are used as healthcare management systems. But the challenge in such a technology care [21] is to access the collected health data or medical history, which may result in poor-quality care. To empower direct interactions among various healthcare stakeholders, BT integrates heterogeneous health database systems with blockchain structure. The high-level conceptual design based on distributed and decentralized ledger technology for patient care is depicted in Figure 5.6.

In the conceptual blockchain-based structure, the middle element with dotted lines represents a secured data channel opened to share patient health data in a distributed technology. The data channel is connected to the keyed file using the smart contract-based system for safe data transactions between the participants.

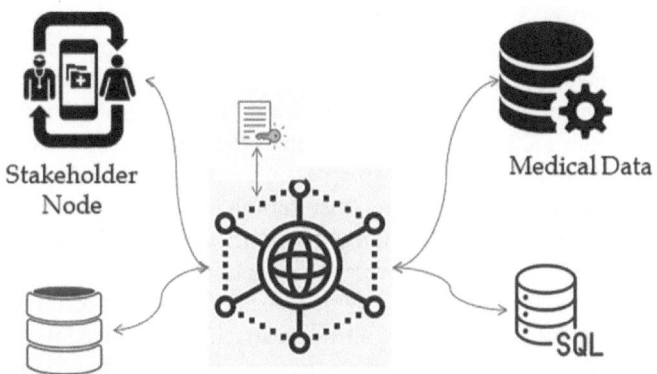

FIGURE 5.6   High-level conceptual design for blockchain-based solution to patient care.

5.6.4   USE CASE FOR CANCER PATIENT CARE

Based on the International Agency for Research on Cancer (IARC) survey report, the latest estimates on the global burden, the cancer burden, is nearly 18 million new cases, and 10 million deaths were reported in 2018. It is essential to reduce the increase of cancer burden globally using effective prevention techniques. It is a challenging factor as it is very complex to diagnose some cancer cases and to identify appropriate treatment options.

Today, a cancer patient may wish to take a second opinion on his or her clinical condition and/or treatment suggestion from a different health professional. In such a situation, the patient and his or her kin needs to carry all the medical documents manually to the health domain expert. In this Internet era, there is a lack of secured data sharing process with the concerned health providers or regional cancer centers/ hospitals and doing so manually is quite time-consuming, which delays the treatment process. To provide a patient-centered care for such cancer patients, BT [21] will provide a trusted relationship between the involving parties. At the same time, blockchain creates a safe network in data sharing features.

According to the study in Ref. [21], data sharing features help us to ensure data integrity, and it decreases the unreasonable repetition of clinical trials. Also, population-based cancer registrations are encouraged across geographical locations for data collection and cancer control.

5.7   HEALTHCARE CHALLENGES AND ITS SOLUTIONS IN BLOCKCHAIN

In the smart communities, there are many smart devices, and mobile applications use digital transactions for communication between various stakeholders. In the healthcare industry, large amounts of medical data are generated. In real-time analysis of medical data set, there is a need of cloud- or fog-based system for efficient results [23].

The paper survey in Ref. [24] specifies the rule-based beacons system using Machine-to-Machine (M2M) messaging proposed for data management. Most of the healthcare networks are managed and controlled by central authority and can lead to failure at certain points. In the healthcare domain, to provide a decentralized and distributed approach, BT is the best option. Blockchain is an emerging key technology to transform or share the sensitive information in trusted and distributed environment. In this section, we will discuss about the various research challenges in healthcare and its solutions.

For healthcare management, there exist certain important concerns over data integrity, security, and privacy. Data integrity specifies the consistency and accuracy of patients' medical data. Privacy refers to private access of individual patient's sensitive data given to the concerned involving parties. While transforming or sharing such medical data, there is chance of intruders attacking sensitive information. For safe data sharing, security is a major concern, especially in healthcare industry.

Hassan et al. [15] explain the various privacy preservation techniques for blockchain-based IoT systems such as wearable gadgets or any wearable health devices using key encryption strategies. Also, privacy preservation techniques in electronic patient records are applied in the healthcare sector to protect the privacy in blockchain-based IoT systems. To protect the patient's identity, various papers discuss the concept of pseudo identities as "proof-of-conformance" for patient registration with healthcare providers.

Omar et al. propose a blockchain-based platform [25] known as MediBchain for pseudonymous healthcare data. It uses MediBchain protocol to guarantee integrity, privacy, accountability, and security of patient medical information.

Interoperability in healthcare is another major concern in the healthcare sector for exchanging patients' health data between different healthcare institutions. In addition, there are numerous challenges in patient-centric interoperability [26] and patient-driven care. Some technical barriers involved in data exchange are agreements made between involving parties, patient matching algorithms, policies and regulations, procedures, and government rules. The major benefits of interoperability in patient care are reduction in duplication, cost, waste, manual data entry, and improvement in overall operational efficiency, and it provides comprehensive clinical care.

5.8 CONCLUSION

In this chapter, various research challenges and their solutions in healthcare sector were discussed. It focuses mainly on the necessity of designing a secured healthcare network model for medical data exchange between various participants. The chapter explores emerging applications in decentralized ledger technology, which addresses the need for health data to be authenticated, distributed, and immutable. Also, various popular applications developed using BT fall into three categories: patient care management, pharmacy supply chain management, and blockchain concepts in genomic markets.

Additionally, stakeholders are considered as an important entity to handle the crucial and sensitive clinical information in patient care management system. The medical data are available either in the form of traditional methods, such as clinical notes, or in digital form, captured using smart wearable devices. For easy accessibility and to provide the interoperability between the heterogeneous hospital systems, a universal standard is defined. The management of health records such as (EMR, her, and PHR are also discussed in this chapter.

To support the study, some use case scenarios in healthcare domain were explained. It depicts the demerits of existing conventional system and how it can be improved using decentralized ledger technology for data sharing among the healthcare providers and insurance companies. Another important field of health domain is telemedicine, which helps to reduce the visitation expenditure for patients based in remote places. It helps a lot to achieve a better patient-centered care in disease diagnosis, evaluation, and treatment procedures by health experts.

Today, with the advent of technologies, the number of cancer patients is growing at a higher rate in the world. Proper cancer diagnosis and treatment plans will help the patients live a better quality of life. Hence, it is clear that BT has a greater impact in healthcare domain to create a secure and distributed network for transparency of digital data transactions. Overall, still a lot of research opportunities are still waiting in the healthcare sector to build a feasible blockchain-based network design.

REFERENCES

1. D. Dave et al. A survey on blockchain technology and its proposed solutions, *3rd International Workshop on Recent Advances on Internet of Things: Technology and Application Approaches (IoT-T&A 2019)* November 4–7, Coimbra, Portugal, Procedia Computer Science 160 (2019) 740–745.
2. https://builtin.com/blockchain/blockchain-healthcare-applications-companies
3. https://hashedhealth.com/
4. A. Sharma et al. *A Study on Blockchain Applications in Healthcare*, Springer Nature Singapore Pvt Ltd., J. C. Hung et al. (Eds.): FC 2018, LNEE 542, pp. 623–628, 2019
5. https://data-flair.training/blogs/blockchain-in-genomics/
6. https://medrec.media.mit.edu/
7. D. Munro, *Data Breaches in Healthcare Totaled Over 112 Million Records in 2015*, vol. 31, Forbes, New York, NY, 2015.
8. C. Lee, Z. Luo, K.Y. Ngiam, M. Zhang, K. Zheng, G. Chen, W.L.J. Yip, Big healthcare data analytics: challenges and applications, in: *Handbook of Large-Scale Distributed Computing in Smart Healthcare*, Springer, Cham, 2017, pp. 11–41.
9. S. Tanwar et al. Blockchain-based electronic healthcare record system for healthcare 4.0 applications, *Journal of Information Security and Applications* 50 (2020) 102407, 2214–2126.
10. I. Mistry et al. Blockchain for 5G-enabled IoT for industrial automation: a systematic review, solutions, and challenges, *Mechanical Systems and Signal Processing* 135 (2020) 106382, October 2019, doi: 10.1016/j.ymssp.2019.106382 0888–3270
11. M. A. Ferrag et al. Blockchain and its role in the internet of things, strategic innovative marketing and tourism, *Springer Proceedings in Business and Economics*, pp. 1029–1038, doi: 10.1007/978-3-030-12453-3_119
12. P. Rakshit et al. *IoT in Healthcare Paradigm, Principles of Internet of Things (IoT) Ecosystem: Insight Paradigm*, http://www.springer.com/series/8578, ISSN 1868-4394, ISBN 978-3-030-33595-3, pp. 263–276.
13. S. Badr et al. Multi-tier Blockchain Framework for IoT-EHRs Systems, *The 9th International Conference on Emerging Ubiquitous Systems and Pervasive Networks (EUSPN 2018), Procedia Computer Science* 141 (2018) 159–166.
14. P. Rathee et al. Introduction to Blockchain and IoT, *Advanced Applications of Blockchain Technology*, Studies in Big Data 60, doi: 10.1007/978-981-13-8775-3_1
15. M. Ul Hassan, M.H. Rehmani, J. Chen Privacy preservation in blockchain based IoT systems: integration issues, prospects, challenges, and future research directions, *Future Generation Computer Systems* 97 (2019) 512–529.
16. S. Xie et al. Blockchain for cloud exchange: a survey, *Computers and Electrical Engineering* 81 (2020) 106526, doi: 10.1016/j.compeleceng.2019.106526
17. T. McGhin et al. Blockchain in healthcare applications: research challenges and opportunities, *Journal of Network and Computer Applications* 135 (2019) 62–75
18. K. Saeedi et al. *Building a Blockchain Application: A Show Case for Healthcare Providers and Insurance Companies*, Springer Nature, pp. 785–800 FTC 2019, AISC 1069, pp. 785–801, 2020. doi: 10.1007/978-3-030-32520-6_57

19. H. C. Lim et al. *Enterprises and Future Disruptive Technological Innovations: Exploring Blockchain Ledger Description Framework (BLDF) for the Design and Development of Blockchain Use Cases*, Springer Nature Switzerland AG 2020, K. Arai, R. Bhatia (Eds.): FICC 2019, LNNS 70, pp. 533–540, 2020. doi: 10.1007/978-3-030-12385-7_39

20. S. Abdullah et al. School of block–review of blockchain for the radiologists, *Academic Radiology* 27 (1) (2020) 47–57

21. P. Zhang et al. Blockchain technology use cases in healthcare, *Advances in Computers* ISSN 0065-2458, Elsevier, Amsterdam, 2018 doi: 10.1016/bs.adcom.2018.03.006

22. V. Vetriselvi et al. Preventing forgeries by securing healthcare data using blockchain technology, *Information and Communication Technology for Sustainable Development*, Advances in Intelligent Systems and Computing 933, pp. 151–158, doi: 10.1007/978-981-13-7166-0_15

23. S.S. Gill et al. Transformative effects of IoT, blockchain and artificial intelligence on cloud computing: evolution, vision, trends and open challenges, *Internet of Things* (2019), doi: 10.1016/j.iot.2019.100118

24. S. Aggarwal et al. Blockchain for smart communities: applications, challenges and opportunities, *Journal of Network and Computer Applications* 144 (2019) 13–48

25. A.A. Omar et al. Privacy-friendly platform for healthcare data in cloud based on blockchain environment, *Future Generation Computer Systems* 95 (2019) 511–521. doi: 10.1016/j.future.2018.12.044

26. W.J. Gordon et al. Blockchain technology for healthcare: facilitating the transition to patient-driven interoperability, *Computational and Structural Biotechnology Journal* 16 (2018) 224–230

6 Improving Security on Blockchain and Its Integration with IoT

M. Kavitha Margret
Department of Computer Science and Engineering,
Sri Krishna College of Technology

E. Golden Julie
Department of Computer Science and Engineering,
Anna University Regional Campus

D. Vijayanandh
Department of Electrical and Electronics Engineering,
Hindusthan College of Engineering and Technology

A. Balamurugan
Department of Computer Science and Engineering,
Sri Krishna College of Technology

CONTENTS

6.1 INTRODUCTION

Blockchain (BC) is a secured information sharing platform that uses distributed ledger technology. Each distributed node is connected together to form a BC. All nodes
are connected to one another with immutable time stamp treated as an open digital
ledger. Distributed nodes are linked to one another in a linear order and form a BC
data structure. Each linked node has pointer reference to refer to the next block in a
BC network. Immutability, decentralization, security, increased capacity, and anonymity are characteristics of BC technology (Figure 6.1).

Each block in a BC has two parts: a header and a data payload. The header stores
the information of each chain length and content; transaction details are stored in
the payload. Hash value (32-bit SHA256) is calculated to give information about the
previous block.

6.1.1 BC Internet of Things

Internet of Things (IoT) connects machines from home or office on a different domain
through Internet in order to collect and control the element to perform some task.

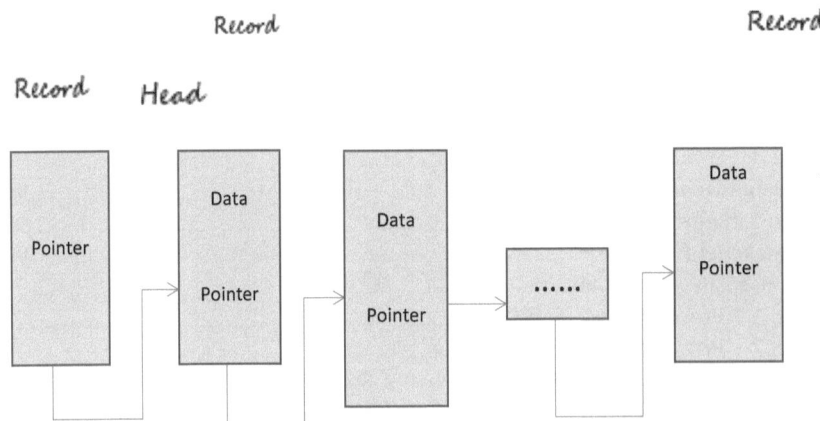

FIGURE 6.1   Structure of blockchain.

Confidentiality, privacy, authentication, denial of service, attacks, self-maintenance, and deployment are the challenges faced in BC integrated with IoT application.

Current investigation works show that BC would be a solution for security alarms regarding IoT [3]. The current IoT network is a centralized system where each network node is managed, and information is authenticated in a centralized manner; scalability is another issue [4] when BC is integrated with IoT. BC offers a decentralized authentication and supervision system that can implement the confidentiality [5], security, and authentication.

6.2 BIOT APPLICATIONS

6.2.1 SMART WEARABLE

Wearable devices are used for computing constraints such as blood pressure, diabetes, blood sugar, obesity, and BMI. As the population is facing such serious health issues, monitoring is needed, which can be done by wearable devices [12]. If the alarming issues are not flagged at emergency times, it will cause big problems for the patient; the output of wearables must be accurate else it will be a very big issue for the patients. Complex algorithms are developed to notice the parametric changes of the human body with fair [13] accuracy. Near-field communication [1] sensors monitor the parameters in the body through both long range and different range communication (Figure 6.2).

Sensors monitor the various parameters in the body by both long-range communication and short-range communication. Collection of data is done by using machine-to-machine communication through collection of networking devices [14]. Wearable devices not only monitor the patients' health parameters but also resources needed in hospitals such as availability of rooms, nurses, equipment, and beds.

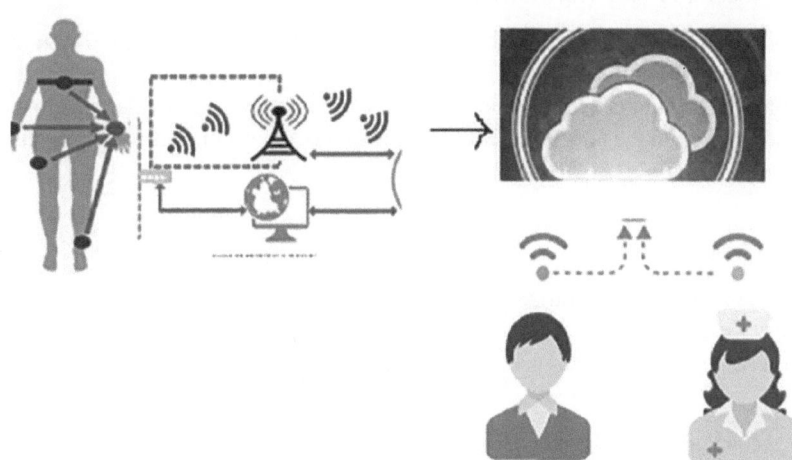

FIGURE 6.2   IOT based health care.

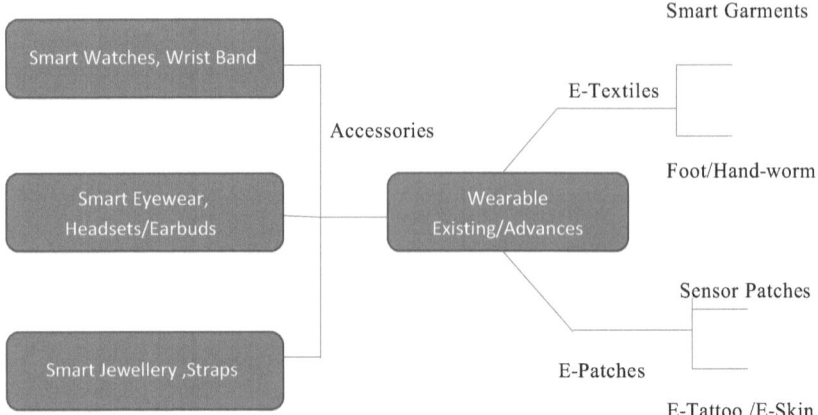

FIGURE 6.3 Classification of wearable devices.

These smart wearables are used for old people who can't be treated in hospitals as they are resistant to hospital zones [15] at elderly age (Figure 6.3).

Blood oxygen can be measured by photodiodes. These diodes measure the quantity of light that is not absorbed, and the variations in the absorbed and nonabsorbed light calculate the blood oxygen [2].

Smart eyewears are special-type contact lenses with sensing mechanism for monitoring eyes. Smart jewellery such as smart rings, clips, and neckwear are used for health monitoring. Armbands and belts monitor physiological signals of humans [4]. Blockchain provides secure transaction of personal health records on the Internet.

6.2.2 SECURITY AND CONFIDENTIALITY OF INDUSTRY

In food processing industry, safety and privacy are very important factors. BC improves security in the industry. IoT's safety and privacy maintaining methods have been proposed by many authors and results have been examined. IoT devices are kept to be secured from the attacker. IBM provides secure environment to ensure security for BC technology.

6.2.3 E-BUSINESS

Smart contract is a decentralized transaction in a BC used in the IoT application and business model proposed by Zhang and wen [9]. Sensor data and user coins can be exchanged without any third party.

6.3 DESIGN GOALS OF BIOT

Decentralization, independence, transparency, security, shared verification, anonymity, and privacy are the design goals of BC.

1. *Decentralization*: BC stores data across the network in a decentralized public. BC eliminates the vulnerability of a single point of failure. It uses public ledger in a decentralized manner. The main advantage of BC is that it eliminates the requirement of a third party as well.
2. *Independence*: BC uses consensus mechanisms where individual entity can control the BC. To exchange the information in the network, BC follows consensus of the majority of the nodes that guarantee reliability and consistency without necessitating any trusted third-party forms [6].
3. *Transparency*: In BC ledger systems, fraudulent transactions cannot be included. It maintains a top level of transparency whenever data are to be updated or entered. It is validated and authenticated by the miners in the network [11].
4. *Security*: A unique cryptographically authenticated signature and consensus mechanism that provides security to each node in the network validates each block. Public key cryptographic algorithm is used by the user to make a digital signature of their transaction. So, the signature will becomes unacceptable if any information is modified in the centralized network.
5. *Shared verification*: Every transaction is verified and added by miners in the BC, which revokes the necessity of a third party in a BC network.
6. *Anonymity*: BC uses a changeable private key (PK) to sign each transaction, and a public key cryptography for authenticating users and controlling access, which provides anonymity.
7. *Privacy*: BC systems offer peer-to-peer privacy and security in decentralized networks. For example, Ethereum have heightened the privacy for BC technology.

6.4 WORKING PROCEDURE

1. *Broadcasting transaction*: Every transaction requested by the user is encrypted using PKs and then broadcasted to the peer networks in the BC.
2. *Validation of Transaction*: Public key cryptographic algorithms can be used to validate the transactions with authentication of the supplicant node.
3. *Block validation*: Every block in the network is validated by executing consensus algorithms such as proof of work (PoW), proof of stake (PoS), or proof of concept in the network.
4. *Appending to BC*: A validated block is included in the BC in a stable and irreversible way.

6.5 BLOCKCHAIN STRUCTURE

BC structure may be different for different applications. Each block has the hash value of the previously generated block, key value, nonce, time stamp, and related information are needed to solve any consensus algorithm such as PoW or PoS puzzle. Key data of the blocks may differ from application to application, for example, IoT machine-to-machine communicated data, smart contract records, bank contract

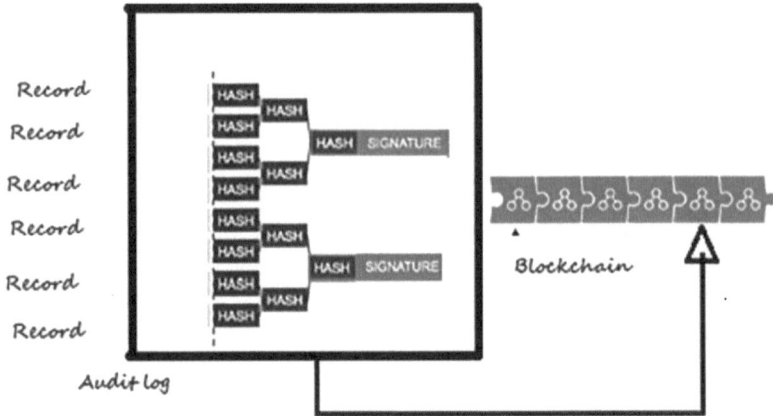

FIGURE 6.4 Protection of database records and audit logs using BC.

records [7], and so on. At the time of each transaction user uses the timestamp as the data proof and it will be sent to receiver node to enable machine to machine communication.

IoT can significantly profit from the techniques provided by BC and will help to develop current IoT technologies (Figure 6.4).

6.6 CONSENSUS ALGORITHMS

To avoid fraudulent activities in BC, the system doesn't rely on third-party authority, rather every node in the network protects against security destructions and keeps track of transactions on nodes to evade indisputable exchange of fraudulent activities.

The consensus mechanism is the agreement among the nodes and shares the common content to be added in every block.

6.6.1 PROOF OF WORK (PoW)

PoW is a defense mechanism against DoS attack [10], Trustworthiness of the data is verified by PoW. It is a technique for finding solution to the mathematical. Lot of computational power needed to solve the mathematical puzzle. Every transaction is verified by the puzzle which is a hash value. The solutions are given by the miners in the network they requires computational power and time to solve the puzzle.

Difficulty level of the puzzles are varied so that it can fix the limit at which new block can be created. Miners will create a conditional junction in the network if more than a one nodes discover a right solution at the same time neighboring nodes who have longest waiting time can accept the blocks in the network.

6.6.2 PROOF OF STAKE

PoW reduces throughput in the entire BC because it requires huge amount of computing power to solve complicated puzzles this can be overcome by PoS mechanism which doesn't require too much computational power.

PoS adds nodes In the BC by paying certain amount of crypto currency which is called as stake. malicious attacks are identified by this node, peer nodes validates a block if it is true block gets added and given as bonus, if the block is rejected in the system, the originator losses some amount of crypto currency the creator losses particular amount of crypto currency. This mechanism will defend against malicious attacks, the origin which initiate attacks loses sake.

6.6.3 Delegated Authenticated Proof of Stake (DaPoS)

Authenticated election process of selection in followed in Delegated authenticated proof of stake.

DaPoS users are involving for election process for "witnesses" and "delegates" with their signs/tokens /keys on the name of their candidate.

Points of the witnesses and delegates may differ in several crypto coins, Each role can engage added role's jobs or even remove it. These dynamic groups may vary by time. Authenticated log by the miners were maintained to check against election process.

6.7 CRYPTOGRAPHIC DIGITAL LEDGER AND REVOCATION LEDGER

A public key infrastructure (PKI) provides protected transportations of information in a network, it is a group of Internet technologies forms a PKI, for example, HTTPS—it assured that it comes from a verified source (Figure 6.5).

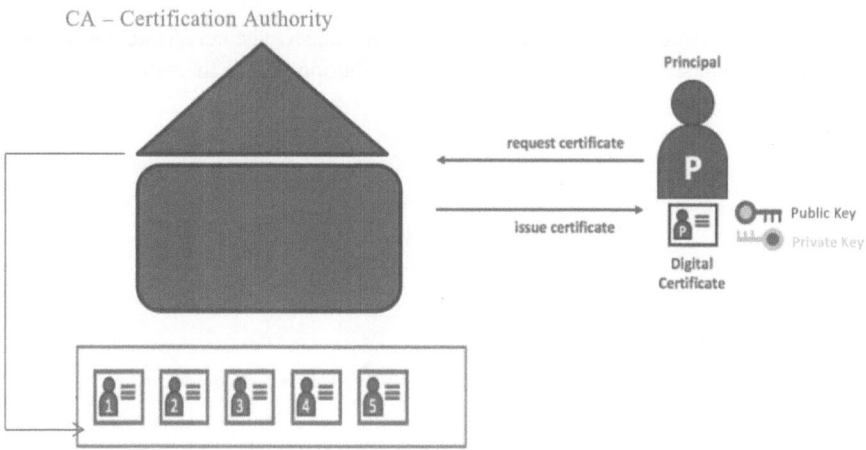

FIGURE 6.5 Public key infrastructure.

The elements of PKI are

1. *Certificate authorities*: CA issues digital certificates to parties and authenticate the parties while exchange with their environment.
2. *CA's certificate revocation list (CRL)*: Revoked certification lists are listed in CRL [8]. It creates a reference for the revoked certificate which are no longer used. Cancelation of a certificate can done for many reasons. For example, a certificate may be revoked because the cryptographic key material has been compromised.

BC network is also relies on the PKI standard to guarantee secure point to point communication between various network, It confirms communication messages posted on the BC are authenticated.

There are four key elements to PKI:

- Digital certificates
- Public and private keys
- Certificate authorities
- Certificate revocation lists

6.8 DIGITAL CERTIFICATES

Set of attributes relating to the owner or user of the certificate is called as digital certificate. A digital is the common type of certificate with X.509 standard. The most common type of certificate is the one compliant with the *X.509 standard*, where encoding of node's finding details is structured I maintained.

Sample X.509 certificate, There are many attributes in an X.509 certificate (Figure 6.6).

A digital certificate describing user placed in subject. The certificate has many of the information such as their public key is distribution within the certificate may be viewed by all, private key must be kept private.

```
Certificate:
    Data:
        Version: 3 (0x2)
        Serial Number: 2 (0x2)
        Signature Algorithm: sha1WithRSAEncryption
        Issuer: O=www.freelan.org, OU=freelan
        Validity
            Not Before: Apr 27 10:54:40 2012 GMT
            Not After : Apr 25 10:54:40 2022 GMT
        Subject: OU=freelan, CN=couchbase-bob/emailAddress=bob@freelan.org
        Subject Public Key Info:
            Public Key Algorithm: rsaEncryption
                Public-Key: (4096 bit)
                Modulus:
                    ...
                    ...
```

FIGURE 6.6   X.509 Certificate

In BC, user attributes can be recorded using cryptographic technique where interfering, alterting will be cancel the certificate. User need to prove her identity using Cryptographic technique allows user to existent her certificate to others in the BCc, Certificate Authority (CA) certificate issuer, keeps cryptographic information, anyone reading the certificate can be confident that the information about user has not been tampered. X.509 certificate is a digital identity card that is terrible to tamper.

6.9 PUBLIC/PRIVATE CRYPTOGRAPHIC ALGORITHMS

Confidentiality and Authentication are needed for secure communication. Authentication ensures that sender and receiver who exchange information are authenticated, A message to ensure "integrity" means that cannot been modified during message transmission. Digital signatures are the traditional authentication mechanisms uses digital signatures and allow the source and destination to sign digitally. This signing mechanism provides guaranteed integrity of the signed message. This mechanisms requires that public key made available to all turns as authentication anchor, and a private key is used to generate digital signatures on the communication.

Receiver can verify the source and trustworthiness of a received message by checking the signature is valid for the sender.

Secure communications is possible a private key and the respective public. With the mathematical connection between the keys private key can be used to produce a signature on the corresponding public key, Both can match, to verify signature if not tampered the source is a valid source (Figure 6.7).

6.10 CERTIFICATE AUTHORITY

Digital identity is required for node to join in the BC network. Trusted authority issues *digital identity*, digital identities are validated forms issued by a CA. More popular CAs are GeoTrust, Symantec (originally Verisign), DigiCert, GoDaddy, and Comodo. These are a common part of Internet security protocols (Figure 6.8).

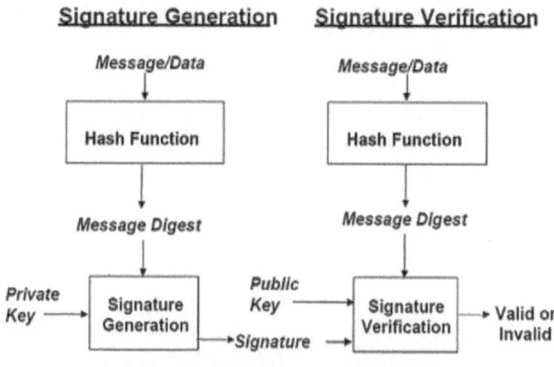

FIGURE 6.7 Model for signature verification.

FIGURE 6.8   Certificate Authority

A CA distributes certificates to different participants in the network. All certificates are digitally signed by the CA and bind together with the actor's public key. CAs also have a certificate, identities of CA can be verified by checking that the certificate have been generated by the holder of the corresponding private key. In BC every node who needs to be a part of network requires an identity. One or more CAs can define the participants in a network from a digital perspective and each participants need to verify their digital identity [16].

6.10.1   Chains of Trust

There are two types of CA

1. *Root CAs*: They *issue* millions of certificates to users
2. *Intermediate CAs*: Have their certificate given by roots CA or another intermediate authority

The above scenario forms creation of a "chain of trust" treated as the best way to track back to the Root CA providing security, delivered certificates to use Intermediate CAs with assurance—if root is compromised the entire chain of trust is in danger (Figure 6.9).

6.10.1   Fabric CA

Fabric provides a built-in CA component, it is a private root CA provides digital identities of Fabric participants those who have of X.509 certificates [17]. Fabric CA manages root or intermediate CA to prove identity.

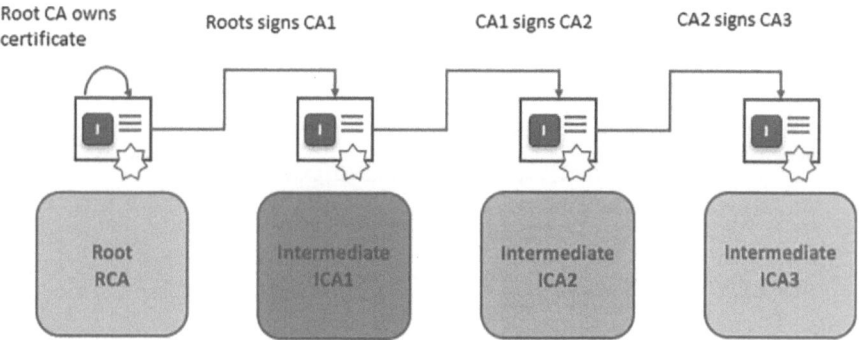

FIGURE 6.9 A chain of trust very helpful in BC system (like Fabric).

6.11 CERTIFICATE REVOCATION LISTS

A CRL is a list with references to certificates that revoked for any reason. When one participant or node wants to verify another nodes identity, it need to check CRL issued by CA, the nodes are treated as a compromised identity. The validity also be checked by CLR. If an actor tries to pass a negotiated or compromised certificate for validating party, it checked against CRL. PKI can provide confirmable uniqueness of node through a chain of trust.

6.11.1 CRYPTOGRAPHY IN BC

Trust in the BC can be developed through consensus and cryptography. Cryptography uses cryptographic algorithms to protect or manage network.

6.11.2 TYPES OF CRYPTOGRAPHY IN BC

Data integrity for BC is validated through Cryptography and hash function.

6.11.3 TYPES OF CRYPTOGRAPHY

- *Public key cryptography*: User authentication and data validation for BC is done by Public key cryptography, authentication is given by public and private key through a PKI. A public key is shared with others and private key stored in private credentials. Any public key algorithms can be used to protect transaction in the BC network.
- *Hash functions*: Cryptographic hashing is a hash code is created by a hash algorithm it takes a digital object and generates a fixed 32-character size string of letters, it is a one-way conversion where the hash is created.
  Hash function generates immutable characteristic of BCs, simple change to the original data will create a new hash, those changes are easily noticeable. Each block in BC have hash code, the code is generated by combination of the plus hash code of the earlier block. Thus chain of hash is created, those

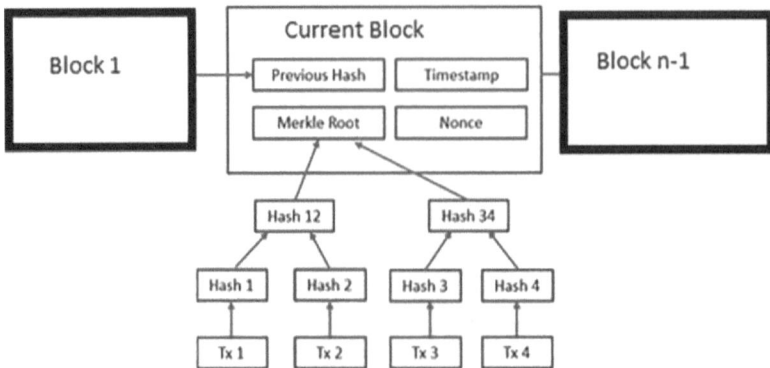

FIGURE 6.10 Merkle tree.

hash code verified but cannot been altered. Any attempt to change the chain of hash codes in will break BC chain of trust and participants.

- *Merkle tree*: Efficient validation of data is done by Merkle on the BC. Each block has a root called Merkle root that is used to stored hash in each block, through this Merkle root data for a current transaction can be sent along with the hash values (Figure 6.10).

To confirm the transaction the authenticating node itself can calculate the hash values to ensure the transaction is valid without validating all nodes in the block.

6.12 CONCLUSION

BC technology is explored in a lot of areas including in public life and governmental affairs the public and private sectors are presently allowing for ways implementing BC technology. For The private the efficiency of BC is high and low cost. In decentralized platform like Business BC technology assurances to determinate many of these issues. In a public platform BC technology is capable to maintain integrity and confidentiality of records stored using fair cryptographic methods but the distributed nature of BCs allows different participants to "own" it. Thus the BC technology accomplishes the basic principle of commercial models of sharing financial prudence and free markets.

REFERENCES

1. K. Christidis and M. Devetsikiotis, "Blockchains and smart contracts for the Internet of Things," *IEEE Access*, vol. 4, pp. 2292–2303, 2016.
2. M. Conoscenti, A. Vetro, and J. C. De Martin, "Blockchain for the Internet of Things: A systematic literature review," in *Computer Systems and Applications (AICCSA), 2016 IEEE/ACS 13th International Conference of. IEEE*, 2016, pp. 1–6.
3. J. C. Song, M. A. Demir, J. J. Prevost, and P. Rad, "Blockchain design for trusted decentralized IoT networks," in 2018 *13th Annual Conference on System of Systems Engineering (SoSE). IEEE*, 2018, pp. 169–174.

4. A. Dorri, S. S. Kanhere, R. Jurdak, and P. Gauravaram, "LSB: A lightweight scalable blockchain for IoT security and privacy," arXiv preprint arXiv:1712.02969, 2017.

5. A. Dorri, M. Steger, S. S. Kanhere, and R. Jurdak, "Blockchain: A distributed solution to automotive security and privacy," *IEEE Communications Magazine*, vol. 55, no. 12, pp. 119–125, 2017.

6. A. Moinet, B. Darties, and J.-L. Baril, "Blockchain based trust & authentication for decentralized sensor networks," arXiv preprint arXiv:1706.01730, 2017.

7. S. Underwood, "Blockchain beyond bitcoin," *Communications of the ACM*, vol. 59, no. 11, pp. 15–17, 2016.

8. T. Hardjono and N. Smith, "Cloud-based commissioning of constrained devices using permissioned blockchains," in *Proceedings of the 2nd ACM International Workshop on IoT Privacy, Trust, and Security*. ACM, 2016, pp. 29–36.

9. Z. Xiong, Y. Zhang, D. Niyato, P. Wang, and Z. Han, "When mobile blockchain meets edge computing: Challenges and applications," arXiv preprint arXiv:1711.05938, 2017.

10. K. Suankaewmanee, D. T. Hoang, D. Niyato, S. Sawadsitang, P. Wang, and Z. Han, "Performance analysis and application of mobile blockchain," in *2018 International Conference on Computing, Networking and Communications (ICNC)*. IEEE, 2018, pp. 642–646.

11. A. Balamurugan and T. Purusothaman, "IPSD: New coverage preserving and connectivity maintenance scheme for improving lifetime of wireless sensor networks," *WSEAS Transactions on Communications*, vol. 11, no. 1, pp. 26–36, 2012.

12. P. T. Selvy, V. Palanisamy, and T. Purusothaman, "Performance analysis of clustering algorithms in brain tumor detection of MR images," *European Journal of Scientific Research*, vol. 62, no. 3, pp. 321–330, 2011.

13. K. Ahmed, F. Ahmed, S. Roy, B. K. Paul, M. N. Aktar, D. Vigneswaran, and M. S. Islam, "Refractive index-based blood components sensing in terahertz spectrum," *IEEE Sensors Journal*, vol. 19, no. 9, pp. 3368–3375, 2019.

14. D. S. Punithavathani and K. Sankaranarayanan, "IPv4/IPv6 transition mechanisms," *European Journal of Scientific Research*, vol. 34, no. 1, pp. 110–124, 2009.

15. S. Sathya Bama, M. S. Irfan Ahmed, and A. Saravanan, "A mathematical approach for improving the performance of the search engine through Web content mining," *Journal of Theoretical and Applied Information Technology*, vol. 60, no. 2, p. 343,350, 2014.

16. N. K. Sreeja and A. Sankar, "Pattern matching based classification using Ant Colony Optimization based feature selection," *Applied Soft Computing Journal*, vol. 31, no. 2818, pp. 91–102, 2015.

17. K. Raveena, K. Elavarasi, and M. Kaaviyapriya, "Survey - web application development," *AJAST*, vol. 2, no. 2, pp. 143–147, 2018.

7 Blockchain-Based Integrated Digital Health Record – A New Model for Health Information Exchanges

Usharani Chelladurai and Seethalakshmi Pandian
UCE, BIT Campus, Anna University

CONTENTS

7.1 INTRODUCTION

Healthcare has always been very essential to people and culture; with the advancement in technology, people have become more conscious about their health. Disease, accidents, and predicaments do arise every day; patients visit different hospitals for consultations with specialized physicians for advanced diagnoses and treatment. Because of this, hospitals have increased, and there have been tremendous changes in the medical industry, with physical health data turning into digital form but patients still moving from hospital to hospital with their physical healthcare records. The main problem is that digital health data are not properly collected, maintained, and networked by the hospitals. A person may suffer from many diseases; he or she

FIGURE 7.1   Integrated healthcare database.

may visit different physicians in different hospitals. In such a scenario, one cannot access his or her health reports quickly outside of the hospital. This leads to many problems, such as (i) prescribe a new drug that will harmfully interact with another medicine you are already having, (ii) requirement of a lab test that one had got done earlier, thus deferring treatment and increasing the expenditure. To avoid new medicines, test reports, and increased expenses, A bitcoin blockchain technology has been facilitated [1]. A blockchain-based integrated digital health record (IDHR) has been generated, which gives digital health data to healthcare providers/doctors rapid, trustworthy and protected access to medical summaries, prescriptions, drug summaries, and test results [2]. In IDHR, we can see remarkable improvement in quality, specifically decrease medical errors, reduction in inpatient stays, and better clinical results [3] (Figure 7.1).

7.1.1   BLOCKCHAIN TECHNOLOGY

Blockchain technology has two components. A block is a record of a transaction or an interaction linked to all existing transactions in the peer-to-peer network, thus creating a chain. Before learning about how blockchain technology works in healthcare, we should learn how blockchain technology simplifies money transfer. In the year 2008, Nakamoto extended the blockchain revolution into cryptocurrency bitcoin. Bitcoin removes the intermediaries and trust parties for money transfer. Bitcoin underlying blockchain technology is private, secure, and reliable [4,5] (Figure 7.2).

7.1.2   NEED OF BLOCKCHAIN TECHNOLOGY IN HEALTHCARE

Blockchain technology has been positively applied in many areas. Bitcoin is the first decentralized cryptocurrency; it is also the first popular blockchain application. After the successful implementation of the cryptocurrency came the era of

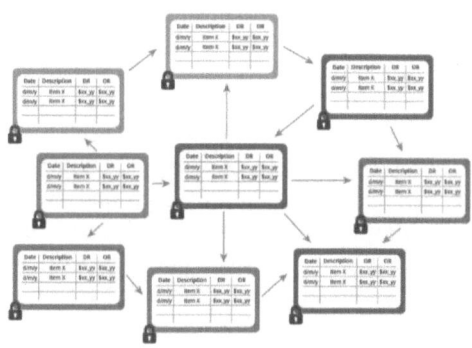

Blockchain is a decentralized P2P architecture. Members in the distributed network record digital transactions into a shared ledger. Each member stores an identical copy of the shared ledger and changes to the shared ledger are reflected in all copies.

FIGURE 7.2 Blockchain P2P network.

blockchain, with the release of Ethereum. Blockchain is a distributed peer-to-peer system that stores all transactions. Traditional databases, such as DDBMS, store huge amount of data, where there is no security. Once data are entered into the blockchain, one cannot change (immutable) the permanently stored data in a digital ledger [6,7]. In a blockchain network, there is no central authority; it distributes the transaction records across all persons in the network [8]. A blockchain-driven health data exchange can reveal the value of interoperability. The health data system operates massive amount of data [9,10]. Exploiting this technology has the possibility to connect uneven systems to generate findings and to better assess the value of care [11]. In future, a countrywide blockchain network for electronic health records may improve competences and support better health outcomes for patients [12,13] (Table 7.1).

TABLE 7.1

Benefits of Blockchain in EHR Management

Blockchain Key Benefits	Improved Electronic Health Record Management
Decentralized management	Patient-managed healthcare records: Patients becomes the owner of their healthcare data
Immutable	Unalterable patient records: The medical data stored off in blockchain cannot be changed by anybody, including physicians and patients
Data provenance	Source-verifiable medical records: Records are signed by source
Robustness/availability	Low risk of patient record storing: The massive amount of medical data is available at anytime and anywhere
Security/privacy	Increased safety of medical records: Data are encrypted in the blockchain and can only be decrypted with the patient's private key

7.1.3 SECURITY AND FRAUD DETECTION IN BLOCKCHAIN TECHNOLOGY

Transfer of digital currency is different from transfer of patient personal data. Massive amount of personal health data are formed through hundreds and thousands of visits. It would be cumbersome for clinicians to search the entire chain and access specific data in a timely manner. Blockchain technology uses numeric public keys and more complex private keys to protect data of each and every individual [14,15]. It is impossible to hack or attack data in the blockchain-driven health system. They cannot find whose data it was or operate the data as they can operate like traditional DDBMS. Likewise, in a blockchain health system, a physician cannot have access rights to all patient records. This sophisticated and intuitive technology addresses all issues that could be an alternative for all current health issues.

7.2 RELATED WORK

In this section, we review some of the existing works of blockchain-based healthcare system.

Nakamoto [1] published the paper titled "Bitcoin", A peer-to-peer blockchain-based electronic cash system. Blockchain is an account book or a distributed database that is stored in a distributed peer-to-peer network without any central control. Once an entry is made in the blockchain-enabled network, it is immutable.

Liu [3] presented reliable blockchain architecture for healthcare services. This system shows a cost-effective insurance billing method. Lan [4] has clearly stated the importance of having a customer feedback mechanism in the e-health system. A feedback mechanism can save manpower and improve the performance of the healthcare system considerably. It gives the pharmaceutical industries a fair chance to produce an improved new drug that is free from any kind of side effects. Liang [5] discusses about including insurance data into an integrated healthcare database. This will ensure that every person in the nation will be able to avail the health insurance, thus making the nation completely covered through a smart health card system.

7.3 INTEGRATED DIGITAL HEALTH RECORD/ DIGITAL HEALTH DATA

The main objective of integrated digital health record (IDHR) is to orient multicultural multilanguage people under one roof. In this IDHR, we can maintain the accuracy of birth to death process of human beings in the plan "One India–One Nation", the country can adopt a national blockchain-based smart healthcare system. The IDHR approach would allow patients to become the owners of their data. The IDHR can be updated as soon as new data are entered anywhere in the country. We can check every moment of the core activity through IDHR throughout the nation. This system can also monitor crude birth rate/death rate: each and every minute we can monitor crude birth rate and crude death rate (Figure 7.3).

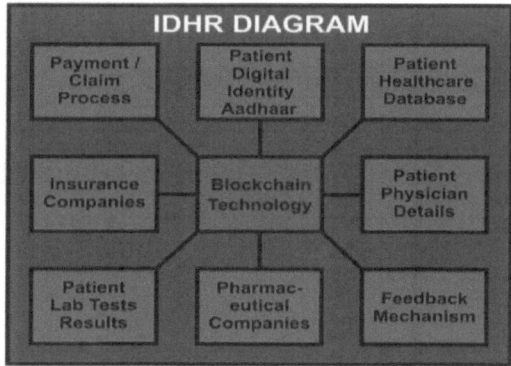

Integrated Digital Health Record

FIGURE 7.3   Integrated Digital Health Record.

7.3.1   OBJECTIVES OF IDHR INTEGRATES DIFFERENT SEGMENTS OF MEDICAL INDUSTRY

- Aadhaar – Digital identity for healthcare services
- Healthcare service providers – Hospitals, doctors, and medical equipment
- Disconnected medical databases across medical industries
- Insurance companies
- Pharmaceutical industries

A. **Aadhaar/digital identity**: "Aadhaar" was created with the objective of issuing a unique identification numbers (UID), named as "Aadhaar", to all residents of India. The system is robust enough to eliminate duplicate and fake identities and can be verified and authenticated in an easy, cost-effective way. In India, 99% of the population is having Aadhaar card, which has been implemented by Government of India necessarily to integrate to the healthcare domain. Integration of Aadhaar brings the original identity to healthcare system to reduce the complexity of personal identification of a patient (Figure 7.4).

   A health record system must have provision to include patient identifiers of the following types (Table 7.2):

   1. Implementers must ensure that the Aadhaar number is available, and then it is a preferred identifier to serve as the unique health identifier. In case the Aadhaar number is not available, the system should allow a user to create more than one identifier, called temporary identifier for each patient in the system.
   2. Identification of patient across the EHR system: Due to lack of mandate for use of Aadhaar or any such alternative(s) national unique identifier, it is difficult to match patient records when exchanging them between two EHR systems. This may lead to situations where different combinations of local identifier and photo identity card numbers of the same person are used at different locations and/or in solutions.

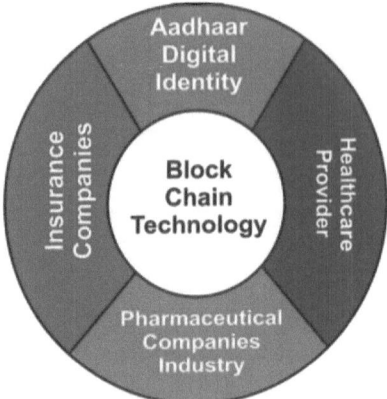

FIGURE 7.4   Integration of healthcare segments.

TABLE 7.2
Aadhaar Details

S.No	Benefits of Aadhaar
1.	Unique ID
2.	Personal details
3.	Irish and thump information
4.	Bank details

B. **Medical databases**: In the current scenario, none of the hospitals is having patients previous medical history. If a patient wants to visit multiple medical institutions for consultation, or is be referred from one hospital to another, there is lag in communication to provide further treatment. Each health service provider has its own software because of which the next physician cannot have a log of the patient's clinical history. The disconnected traditional databases are not enough for predicting upcoming lifestyle and life threat diseases. One solution is that IDHR facilitates to integrate the disconnected medical databases, brings the birth to death clinical summary and all medical data for faster treatment (Table 7.3).

C. **Insurance companies**: In India, there are several health insurance companies that have enormous health insurance policies to cover the cost of medical treatments. In India, all citizens receive some health coverage from their governments, paid for by taxation. Health insurance is often part of an employer's benefits. Most people are not aware about insurance because of their economic condition, illiteracy, and hesitation to enquire about. Few people have government-sponsored health insurance policies, but they are not able to properly use these on time due to one reason or the other, which

TABLE 7.3

Comparison of Blockchain-Based IDHR and Traditional Medical Database Systems

Blockchain-Based IDHR	Health Records in Traditional Database
Decentralized management	No decentralization/anyone can manage
Patient managed	Failure of one will affect the entire system
Immutable/unaltered	Easy to alter the data
Data provenance	Anyone can sign in and access the records
Source-verifiable medical records	
Robustness/availability	High risk of patient record storing: Due to unavailability
The massive amount of medical data are available at anytime and anywhere	
Security/privacy	Less security and privacy of medical records:
More security and more privacy	Because everyone gets access permission

leads to an increase in private health insurance companies. Generally, health policy covers whole or part of the risk of a person incurring medical expenses. Some policies do not cover life threat diseases. In India, senior citizens are the major revenue generators for the medical industry. Bringing those insurance and health records to hospital is a bottleneck to the patients. In that situation, integration of healthcare and insurance is important to find the right policy, which reduces the costs by a fair margin; directly removing the overhead cost associated with the middlemen and intermediaries by integrating insurance companies and the data regarding the medical expenditure and payment of the medical bills can also be added, which would lead to a perfectly chained and transparent healthcare service across the nation (Figures 7.5–7.7; Table 7.4).

Health Insurances in Urban 18% and Rural 14.1%

- Insurance Companies in India
- Government Health Insurance
- Exclusive Health Policies by Corporate
- Other Than Life Insurance

FIGURE 7.5 Insurance companies in India.

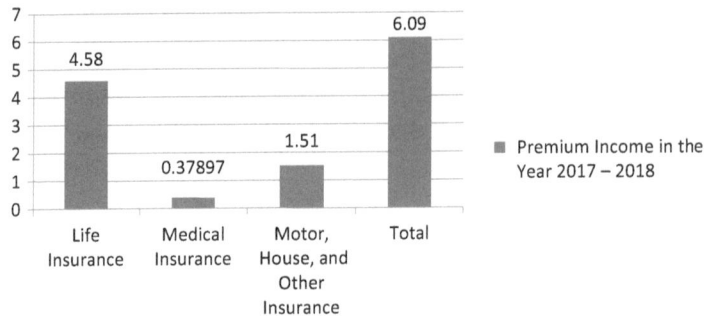

FIGURE 7.6   Approximate premium income 2017–2018.

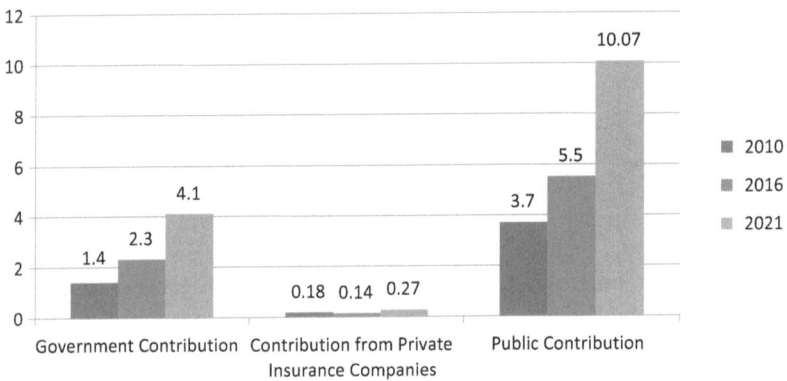

FIGURE 7.7   Insurance companies' contribution growth 2016–2021.

TABLE 7.4
Benefits of Blockchain-Enhanced Claim Process

Blockchain Key Benefits	Enhanced Insurance Claim Process
Decentralized management	Real-time claim processing: The capability to remove intermediaries from insurance claim process is the ability that sets blockchain apart from other technologies
Immutable	Improved claim auditing and fraud detection: All insurance payer, private and government insurers, and individual payers have the benefits of audits facilitation and better fraud detection
Data provenance	Verifiable records for claim qualification
Robustness/availability	Enhanced accessibility of patient data: patient data accessible from multiple providers
Security/privacy	Increased safety of medical records: Increased security of patient medical insurance information

D. **Pharmaceutical industry**: Pharmaceutical companies play a vital role in the healthcare domain. They invest tremendous amounts of money in research, development, and marketing. India is the largest provider of generic drugs globally. The Indian pharmaceutical industry supplies over 50% of global demand for various vaccines, 40% of generic demand in the United States, and 25% of all medicine in the United Kingdom. Indian pharmaceutical companies play an important role in the global pharmaceutical sector. The country also has a large pool of scientists and engineers who have the potential to steer the industry ahead to an even higher level. Presently, over 80% of the antiretroviral drugs used globally to combat AIDS (Acquired Immune Deficiency Syndrome) are supplied by Indian pharmaceutical firms. But still the industry is isolated from patients. It is necessary to integrate pharmaceutical data into the smart healthcare domain through a feedback mechanism that supports integration among patients; physician and health service providers who can improve pharmaceutical products and physicians can easily identify resistance of medicine and allergy.

Drug resistance, allergy, and fake medicines can be identified only through blockchain-based IDHR. Without integration of medical databases among health service providers, physicians will not be able to identify drug resistance and allergy to patients. This will create negative results to patients. For life threat/style diseases, drug resistance is taking vital role in the health domain.

7.3.2 Pharmaceutical Industry's Revenue

Indian pharmaceutical companies are capitalizing more on export opportunities in regulated and semiregulated markets (Figure 7.8). In financial year 2017, India exported pharmaceutical products worth around US$ 16.8 billion, with the number expected to reach US$ 40 billion by 2020. During April 2017–February 2018, India exported pharmaceutical products worth Rs. 767.17 billion (US$ 11.90 billion).

Pharmaceutical Industries' Growth over the Period 2010–2020

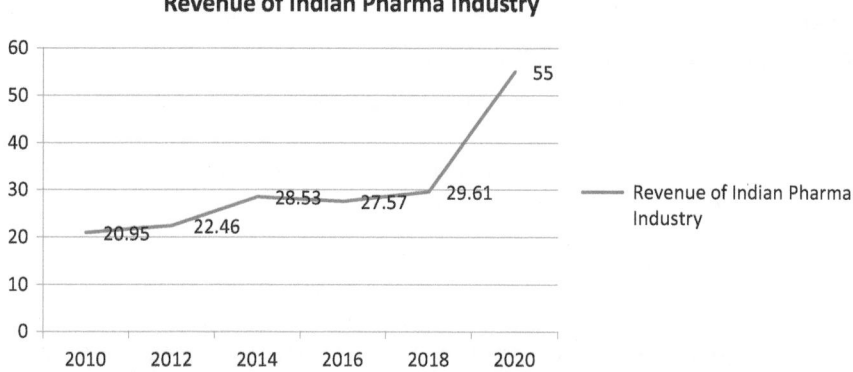

Revenue of Indian Pharma Industry

FIGURE 7.8   Revenue of Indian pharma industry 2010–2020.

- India is the world's largest provider of generic medicines; the country's generic drugs account for 20% of global generic drug exports (in terms of volumes).
- Indian drugs are exported to more than 200 countries around the world, with the United States being the key market. Around 40.6% of India's US$ 16.8 billion pharmaceutical exports in 2016–17 were to the American continent, followed by a 19.7% to Europe, 19.1% to Africa, and 18.8% to Asian countries.

7.4 RESULTS

The Blockchain-Based IDHR System Aims to Provide
- Patient log/patient treatment history
- Improved communication between *health* professionals
- Improved security and confidentiality of patient data and records
- Reduce duplication of data and procedures/examinations, prescription, or referrals
- Simplify claim processing/payment through insurance
- Monitor the pharmaceutical supply chain/quality of drug/prohibited medicines
- Improved clinical decision making and physicians responsibilities
- Avoid repeated laboratory test facilitate fast treatment
- Remove fake doctors/medicines
- Pattern generation disease/medicine
- Disease prediction lifestyle diseases and life threat diseases
- Fraud detection improper treatment identification
- Ensure better quality in healthcare

Table 7.5 shows a comparison of healthcare measures with both blockchain-based IDHR and traditional database systems (Figure 7.9; Table 7.5).

TABLE 7.5
Blockchain-Based IDHR Quality Metrics

S.No	IDHR Quality Metrics
1.	Population sex ratio at birth
2.	Recommending the necessary vaccinations such as polio, rheumatic fever, tuberculosis
3.	Blood grouping/rare blood grouping identification
4.	Adulteration
5.	Maternal mortality ratio
6.	Keep track of seasonal diseases such as malaria, dengue, plague, swing flu
7.	Supportive for national mission: Population control (family planning)
8.	Physicians registered with MCI (Medical Council of India), restrict and identify fake doctors
9.	Monitor major disease ratio and disease-affected regions
10.	Reduce death rate and improve quality outcomes in overall health system

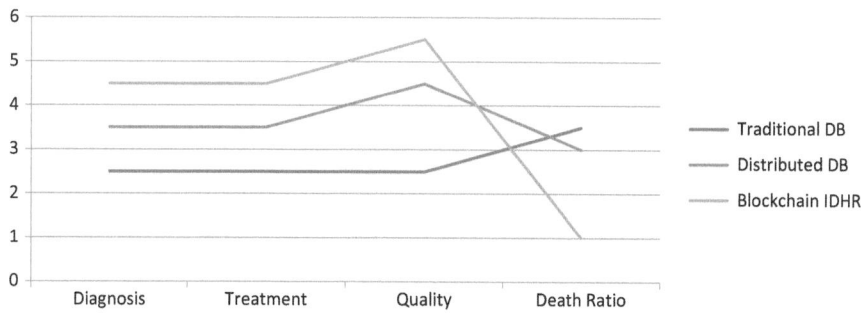

FIGURE 7.9 Performance measures compared with existing traditional health systems.

7.5 CONCLUSION

The proposed blockchain based Health Information Model, which provide decentralized management, an immutable audit trail, data provenance, robustness/availability, and security/privacyThe proposed model has discussed the importance of blockchain-based healthcare system, which integrates patients with personal data, electronic health data, hospitals, physicians, insurance companies, drug industry to improve health record management, enhanced insurance claim process, and accelerated clinical/biomedical research. The smart healthcare system adopts an integrated healthcare model for processing the personal health data of individuals through a blockchain network, thus ensuring the data ownership of individuals as well as transparency, integrity, confidentiality, speed, scalability, threat, and attack. However, these challenges can be addressed through careful application, design, and implementation; therefore, healthcare applications of blockchain continue to increase. Blockchain-distributed ledger technology can advance the biomedical and healthcare domains in various novel ways, and we expect many new applications to emerge soon.

This system also helps to achieve ONE INDIA–ONE HEALTH CARD.

REFERENCES

1. Satoshi Nakamoto (2008): "Bitcoin" *A Peer-to-Peer Electronic Cash System*, satoshin@ gmx.com www.bitcoin.org.
2. Dixon, Brian E., and Caitlin M. Cusack. (2016). Measuring the value of *health information exchange*. In Health Information Exchange Edited by Brian E. Dixon (pp. 231–248). Apple Academic Press.
3. Liu, W., Zhu, S. S., Mundie, T., & Krieger, U. (2017, October). Advanced block-chain architecture for e-health systems. In 2017 *IEEE 19th International Conference on e-Health Networking, Applications and Services* (Healthcom) (pp. 1–6). IEEE.
4. Duan, Y., Zhou, G., Zhang, Y., Lan, Z., Chi, C., & Yan, W. (2018, June). Neural Network Based Clinical Treatment Decision Support System for Co-existing Medical Conditions. In 2018 *IEEE Third International Conference on Data Science in Cyberspace* (DSC) (pp. 131–138). IEEE.
5. Liang, X., Zhao, J., Shetty, S., Liu, J., & Li, D. (2017, October). Integrating blockchain for data sharing and collaboration in mobile healthcare applications. In 2017 *IEEE 28th annual international symposium on personal, indoor, and mobile radio communications (PIMRC)* (pp. 1–5). IEEE.

6. Mettler, M. (2016, September). Blockchain technology in healthcare: The revolution starts here. In 2016 *IEEE 18th international conference on e-health networking, applications and services (Healthcom)* (pp. 1–3). IEEE.

7. Zhou, J., Cao, Z., Dong, X., & Lin, X. (2015, April). TR-MABE: White-box traceable and revocable multi-authority attribute-based encryption and its applications to multi-level privacy-preserving e-healthcare cloud computing systems. In 2015 *IEEE Conference on Computer Communications (INFOCOM)* (pp. 2398-2406). IEEE.

8. Mohammad Nabil Almunawar, Muhammad Anshari (2014): Empowering customers in electronic health (e-health) through social customer relationship management, *Int. J. Electronic Customer Relationship Management*, vol. 8, 44–57.

9. Shan Jiang, Jiannong Cao, Hanqing Wu, Yanni Yang, Mingyu Ma, Jianfei He (2018): BlocHIE: A Blockchain-Based Platform for Healthcare Information Exchange, In 2018 *IEEE International Conference on Smart Computing (Smartcomp)* (pp. 49–56). IEEE.

10. Leslie Mertz (2018): (Block) Chain Reaction: A Blockchain Revolution Sweeps into Health Care, Offering the Possibility for a Much-Needed Data Solution, *IEEE PULSE*, vol. 13, 4–7, May/June.

11. Hassan Qudrat-Ullah, Peter Tsasis (2017): *Innovative Healthcare Systems-An Introduction, Innovative Healthcare Systems for the 21st Century*, Springer International Publishing, ISBN: 978-3-319-55773-1, 978-3-319-55774-8

12. X. Xu et al. (2016): The Blockchain as a Software Connector, In 2016 13th Working IEEE/IFIP Conference on Software Architecture (WICSA) (pp. 182-191). IEEE.

13. D. Koutsouris, A. Lazakidou (2014): *Concepts and Trends in Healthcare Information Systems* (Vol. 16). Springer.

14. Arni Ariani (2017): Innovative Healthcare Applications for Developing Countries, *Innovative Healthcare Systems for the 21st Century*, Springer, 2017, ISBN: 978-3-319-55773-1, 978-3-319-55774-8.

15. Charles A. Shoniregun, Kudakwashe Dube, Fredrick Mtenzi (2010): Framework Securing e-Healthcare Information, *Electronic Healthcare Information Security* (Vol. 53). Springer Science & Business Media.

8 A Complete Study on the Major Protocols of Blockchain

S. Palanikumar and Sivaprasad Abirami
Noorul Islam Centre for Higher Education Shah
and Anchor Kutchhi Engineering College

CONTENTS

8.1 INTRODUCTION

The following are the major problems with bank transactions:

- The technical issues
- The account might be hacked
- Exceeded transfer limits
- High transfer charges

The above problems can be solved by cryptocurrency.

Before we start with the introduction of blockchain, it's very important to understand the difference between two major terms: bitcoin and blockchain.

Bitcoin: It is a cryptocurrency that is established globally to store values and was started in 2009[1–3].

Blockchain: It is a backbone technology adapted by the digital cryptocurrency for secured transactions. Blockchain technology was commercially implemented first by bitcoin. Blockchain was introduced in 2008[1–3].

Blockchain provides a decentralized architecture on a peer-to-peer network. It is a ledger of transaction shared among multiple peers on a distributed environment to avoid security flaws. All peers participate in the transaction and inspect the whole process instead of having a single user control that ensures authentication and avoids any hacker to mislead the transaction with the help of public key cryptography and the hash function [1,3].

In recent years, blockchain has been adapted in varies fields such as medical, financial, government, and many more [4].

8.2 BLOCKCHAIN BASICS

Let us understand the basic elements of blockchain, which is illustrated in Figure 8.1.

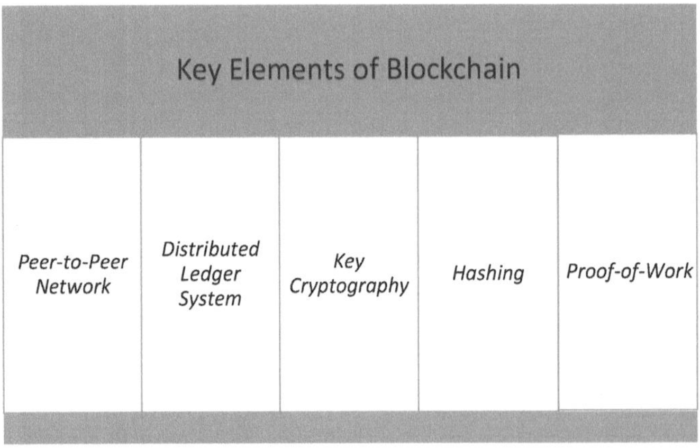

FIGURE 8.1 Key elements of blockchain.

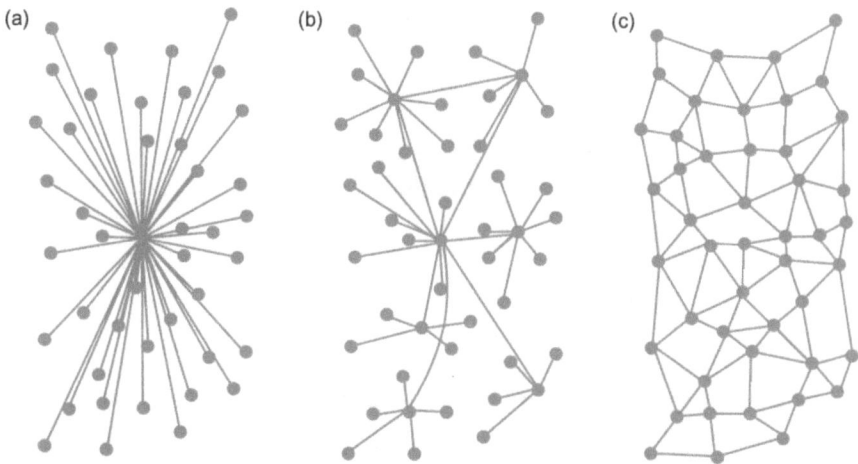

FIGURE 8.2 Various network models. (a) Centralized network, (b) decentralized network, and (c) distributed network.

8.2.1 PEER-TO-PEER NETWORK (P2P)

This network is distributive in nature and is in a decentralized manner. There is no concept of centralized service, instead all the nodes are linked with each other to make the transaction or data transfer (Figure 8.2) [5].

8.2.2 DISTRIBUTED LEDGER SYSTEM

In the distributed ledger system, all nodes will keep a copy of the data record instead like a centralized system where in the single centralized node keeps the data record. All nodes participate in the data transfer or transaction, for which all nodes carry the updated copies of the ledger or the data record in the blockchain network. All the participating nodes of the transaction will have the same copy of the data record and the record of each other node. As everyone has the updated copy of the data record, no one can manipulate/falsify the transaction.

8.2.3 KEY CRYPTOGRAPHY

During the transaction in a blockchain, the confidentiality, authentication, and integrity are provided with the help of an asymmetric key cryptography. In asymmetric key cryptography, two keys are used: the private key, which is used to encrypt the data/transaction that is available with the intended user to do the transaction, and the public key, which is distributed over all the participants using the elliptic curve Diffie–Hellman–Merkle and is used to decrypt the data/transaction. The combination of private and public key ensures the data security.

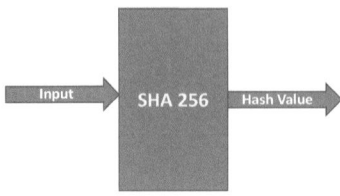

FIGURE 8.3   Hashing techniques.

8.2.4 HASHING

It is a method of nonreversible encryption used in blockchain to improve the transaction security. Hashing converts the variable length data into fixed length data as output using the secure hash algorithm 256 (SHA 256) (Figure 8.3).

8.2.5 PROOF OF WORK

The proof of work from each node of the decentralized peer-to-peer network verifies the transaction. All the participants are supposed to do complex computations to validate themselves on the network as a valid participant. The computational power of the legitimate participant is to be greater than the hacker computational power to make a legal transaction. The Sybil attack and the double spending problems that are discussed in the later part are removed with the help of proof of work.

The proof of work contains a complex mathematical cryptographic calculation that is identical to Adam Back's hash cash [7]. The proof of work scan is a special value called nonce (number only used once), which is a number added to the hashed value in a blockchain with the help of SHA-256, which results in a hash that starts with many zero bits. The target will set the required number of zero bits. The output hash contains a number of zeros in the beginning, which should be less than that of the target. The proof of work is verified by incrementing the nonce to achieve the target number of zeros bits in the starting of the hash generated by the corresponding block. Once the computation is completed, it cannot be changed. If the hacker tries to change the computation, then the verification will be done by all the other blocks involved in the transaction since all will be having a different hash value than the hacked value. If the hacker wishes to change the block, then he or she should have a very high computational power to change the hash value in all the blocks that participate in the transaction (Figure 8.4).

8.3   BLOCKCHAIN PROTOCOLS

Blockchain is a growing and popular technology. After the growth and spread of bitcoin protocol, many other protocols have been designed for various applications in different domains. In this chapter, the major protocols are explained with the features, the transaction method, and the advantages and the disadvantages. The major protocols are illustrated in Figure 8.5.

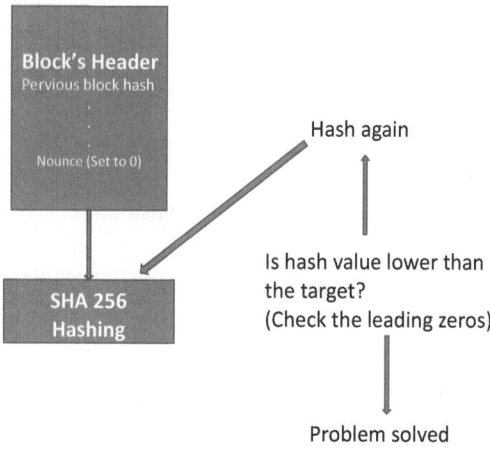

FIGURE 8.4 Proof of work.

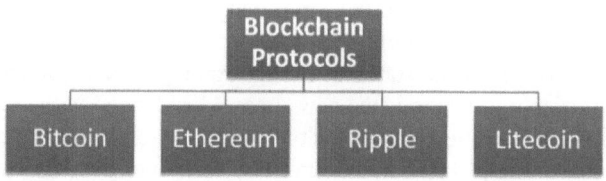

FIGURE 8.5 Blockchain protocols.

8.3.1 BITCOIN PROTOCOL

8.3.1.1 What Is Bitcoin?

A decentralized peer-to-peer electronic/virtual cash payment system is the bitcoin. Bitcoin is the first popular cryptocurrency and a global payment system. Satoshi Nakamoto created bitcoin in the year 2008. The technology used for the transaction process of bitcoin is blockchain. As discussed earlier, blockchain has been adopted in various domains in various industries for the transaction of cryptocurrency. At the end of September 2019, the size of the bitcoin blockchain is around 242.39 GB [6].

8.3.1.2 Why Bitcoin?

Clashes with a third party – bank in the traditional system – creates crisis in a transaction. The high transaction cost in the traditional system reduces transactions, because the third party charges high transaction fee and other charges. The next major issue of the traditional system is security. The traditional system is easily hacked by hackers, which creates a privacy problem. Nakamoto focused on the above issues and created a secured payment system that is based on the decentralized peer-to-peer system with cryptographic features.

8.3.1.3   Features of Bitcoin

The following are the features of bitcoin:

- Decentralized peer-to-peer digital network for the payment transaction.
- No involvement of third party, results in the removal of third party trust verification.
- The security process used is irreversible, which protects the online transaction from hackers.
- The timestamp system used by the bitcoin guarantees the sequential order of the transaction.

8.3.1.4   Transactions

Nakamoto proposed a bitcoin with a P2P decentralized architecture with a time stamp server that makes the order of the transaction in a proper way [9,10]. A cryptocurrency is a chain of digital signatures, and each transaction is a set of digitally signed hash value of the descended transaction and the public key of the next block. The purpose of private and public key combination is to provide an asymmetric key cryptography. The private key is used by the initiator of the transaction to encrypt the transaction, and the public key is used by all other participants to decrypt and verify the transaction (Figure 8.6).

The state of the transaction defined in the bitcoin is called the bitcoin ledger. The bitcoin ledger consists of the status of all the existing bitcoin ownership statuses. The output of each transaction is the new state and is defined in the state transition function. The state transition is accepted only when the sender of the transaction has enough bitcoins to complete the transaction, else the system generates an error. Each transaction is a set of input and output that is uniquely identified by its hash value. The output of each transaction can be used only once as the next transaction input in the whole blockchain process [10,11].

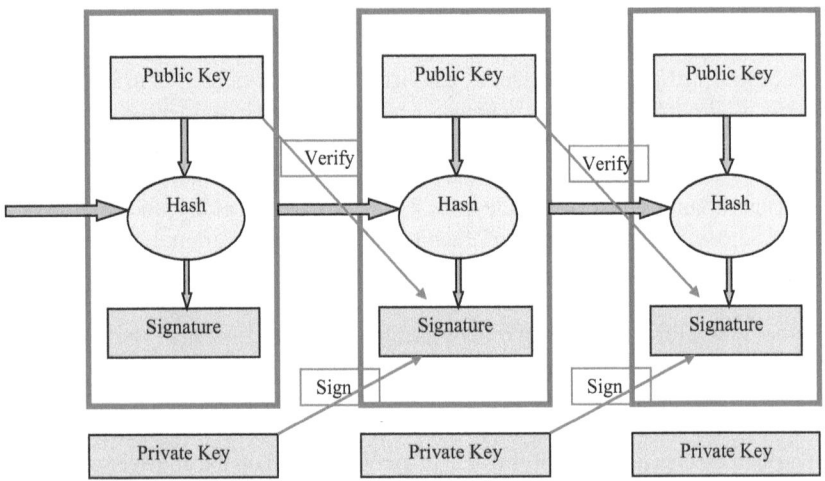

FIGURE 8.6   Bitcoin logo.

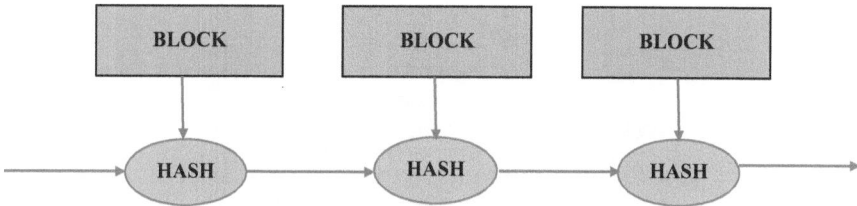

FIGURE 8.7   Bitcoin transaction [9,10].

Double-spending problem: This occurs when the same cryptocurrency is used more than once. It can also be defined as in any bitcoin transaction where the same input is used for multiple transactions that was used previously and got broadcasted on the decentralized network [10].

Unspent transaction output (UTXO): It is defined as an output of the transaction that is not used or spent in the next new transaction. This UTXO is used mostly in all the cryptocurrencies [9,10].

Spent transaction output (STXO): It is define as an output of the transaction that is used or spent in the next new transaction [9,10].

In any bitcoin transaction, it can have maximum of two outputs and multiple inputs.

Output: It is the amount sent to the other participant or the balance that is returned back to the sender.

Input: The multi-inputs can be utilized to join smaller amounts of coins transferred.

Sybil attack: This is a type of attack common in a peer-to-peer network to strengthen the power in the reputation system. The purpose of this attack is to achieve the power in the network to do illegal transaction in the network. A single partici- pant builds numerous identities/accounts to do the banned transaction, but for the other participants, the different fake accounts/identities seem to be genuine unique accounts/identities.

All transactions and ownerships are described clearly in the distributed ledger used in the bitcoin network. All participants of the decentralized peer-to-peer net- work have a copy of the bitcoin ledger [12]. When any participant wishes to do a transaction to send coin to another participant, the transaction should be informed publicly, and the decentralized peer-to-peer network will take care of the transaction with proper verification and correctness. But in the network, the participant can try to change the network and try the same coin transaction with two different partici- pants, which is already discussed earlier and called as double spending problem. The same participant can arrange multiple instances of the transaction to verify the initial transaction, which leads to Sybil attack. The problems discussed above are solved using proof of work in the bitcoin network.

The hash is generated and announced publicly by the time stamp server that con- firms the existence of the data inside the block during the time of hashing. The time stamp server also verifies the time stamp of the current and the previous blocks; the

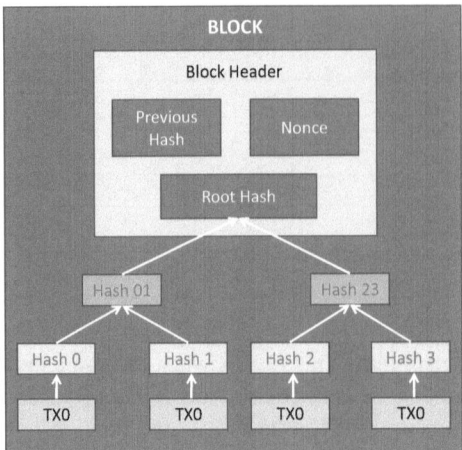

FIGURE 8.8 The Hash chain with the block [10].

time stamp of the current block should be greater than that of the previous block. The blockchain links the hash of the blocks as shown in Figure 8.8, and the most useful nature of the blockchain is that the history of the transaction can be traced at any time.

The Merkle tree is a binary tree, also called as Hash tree, that is used for data verification. It provides data integrity and maintainability. Each nonleaf node of the tree has the hash value of its child node. The root node shows the hash value of the entire tree. The level of the tree is maintained the same [13,14]. Any change by the hacker in the tree will get reflected in the blockchain [12]. Once the transaction is complete and the verification of the block is done, all hashes in the tree are discarded by the network other than the root hash that gets added in the block header. Bitcoin established a simplified payment verification (SPV) system, which uses the copy of the block header instead of the full transaction history [9,10].

Bitcoin mining: The creator of the block initiates the first transaction by generating a new coin; this triggers the other nodes to start the verification and circulate the coin, which in turn is called as the coin base transaction. All the participants of the transaction are to be honest enough to do the verification as there is no central authority to do the verification process. The when the network initiates a new transaction, it is broadcasted to all the participants. Each participant of the transaction works on the proof of work, then it forward its block to the bitcoin network. The participant in the bitcoin network validates all the transactions before accepting the block. Once the block is validated and accepted, the next block is created, which contains the hash value of the previous block and continues to form the chain. Along with the block creation, the nodes are assigned with coins by verifying the bitcoin transaction. Mining is defined as the process of creating the blockchain by adding new blocks [8–10].

Bitcoin fork: The inconsistent state of the bitcoin network leads to fork; this situation arises when several nodes simultaneously broadcast the same block as the network is distributed and decentralized in nature. There is a possibility of several blockchains getting created from different blocks. This situation is managed by considering the network with the longest chain. Meanwhile there will be an agreement within the network regarding the correct chain and the blockchain generated by the fork will be made invalid [8,10].

One of the major problems of bitcoin is the scalability problem. To achieve the bitcoin network with 1 MB block size, with the transaction capacity of 300 byes, the estimated throughput of bitcoin is 8 GB per 10 minutes. This shows that only the node with the highest computational power can participate in the blockchain, and this is against the purpose of blockchain and bitcoin. Several solutions are tried, which indeed result in different types of forks.

One of the solutions for the scalability problem of the bitcoin is the Segregated Witness (SegWit), which deals with the transaction flexibility. The problem is due to the reality as the transaction signature never covers all the transaction data that give way to the hacker node to modify the transaction and its hash [15]. The block size of the bitcoin is increased along with the new layer that is created on top of the network, which segregates the transaction data from the signature data that felicitate the lightning network [16]. The lightning layer was activated in August 2017 [17]. In this lightning network, a new facility of microchannel is created to make the micropayments in the network. This payment channel acts as an agreement between the sender and receiver to announce a delay in the transaction but make the actual transaction. Both the parties delay sending the transaction details on the network but both assure their balance on the bitcoin blockchain. The victorious examination of the lightning network of the bitcoin was conducted in January 2018 [18].

Advantages of Bitcoins [19]

- Payment freedom

    The participant can send and receive money at anytime anywhere. There are no restrictions as bank holidays in the traditional way.
- Control and security

    The transaction in the bitcoin doesn't now involve any personal data and bitcoin defend from identity theft.
- Information is transparent

    The transaction can be traced back to verify the legitimate transaction and the distributed ledger that is available; public contains all the transaction information, which makes it hard for any hacker to get involved in the transaction.
- Very low fees

    No extra fees of the merchant can be added without the prior confirmation of the participant.

Disadvantages of Bitcoins [19]

- Lack of awareness and understanding

    People are not still aware of the digital currency; and so it is not as popular as the physical currency. Some kind of awareness has to be created through advertising, awareness camp, and lectures.

- Risk and volatility

    The risk of using bitcoin is high; once the bitcoin wallet is lost or corrupted, it cannot be recovered. The coins will be donated to the system, which makes any rich bitcoin user to become poor within a second.

- Still developing

    As the system is still in the developing stage, still there is no proper refund strategy described as the value of the bitcoin changes constantly according to the current requirement.

8.3.2   ETHEREUM PROTOCOL

8.3.2.1   What Is Ethereum?

It is the world's foremost globally recognized programmable blockchain that offers an open-source platform for decentralized application. The digital values are controlled by writing codes that run as per the code and can be accessible in any part of the world. Everyone has the freedom to create the blockchain by modifying the code that operates on the Ethereum environment. The Ethereum is the second largest blockchain protocol that was introduced in 2015 by Vitalik Buterin [20,21].

8.3.2.2   Why Ethereum?

Like bitcoin, Ethereum also overcome the disadvantages of the traditional payment system and also overcome disadvantages with the centralized system. Apart from these, Ethereum is the largest decentralized application (Dapps) that creates its own blockchain. It helps in creating the smart contracts and various decentralized applications without any third-party involvement and downtime. Unlike bitcoin, which is working only in the financial sector, Ethereum is utilized in various sectors, and it also allows the programmers to develop and bring out various distributed applications [22,23].

Features of Ethereum
- Ethereum develops a computing platform that cannot be modified or stopped.
- Ethereum has large passionate community building blocks and distributed applications.
- Elliptical cryptography is used for wallets, proof of stakes (PoS) to secure blockchain.
- Provides a fast block generation for quick transactions and fast scalable programs.
- It brings the concept of private Ethereum networks to the entire world.

8.3.2.3   Transaction
Proof of Stakes

This is the agreement algorithm used in the public blockchain that basically depends on the validator's stake in the Ethereum network. PoS are more energy efficient as compared to PoW, which is used as the agreement algorithm in the bitcoin. In this agreement, a set of validators propose and vote to create new block; the stake of the validator's vote depends on the deposit size. The agreement algorithm can assign the rewards of the validator, which can have multiple logics based on the stake of the validator on the network. The major advantages of PoS are that they have high security, they are more energy efficient, and they remove the problems of the centralized network.

Ether

It is a token value used in the Ethereum network, which is denoted as "ETH" during the exchange of cryptocurrency at the time of making the payment for transaction fees, computational fees, and for gas.

Gas

It acts as an intermediate token during the transaction to calculate the computational power needed to run the smart contract of this transaction. Before the payment is done by Ether, the gas measure the unit of the computational power and based on that the Ether is paid for this transaction (Figure 8.10).

Smart Contracts

The smart contract is the agreement done in the blockchain network to do the transaction to exchange data or money. This third-party application agent mentions the legal terms and conditions that have to be followed during the transaction.

Each block contains an address and state transition that is held in the Ethereum state. The Ethereum state maps the address and the states of the account [10,25]. Two types of accounts are supported by Ethereum:

- Externally owned account (EOA), which is controlled by private key
- Contract accounts, which are controlled by contract code [12]

The four basic fields of the Ethereum account are balance of Ether, contract hash code, nonce, and storage root [10,25,26].

Ether: (Discussed earlier in the chapter)

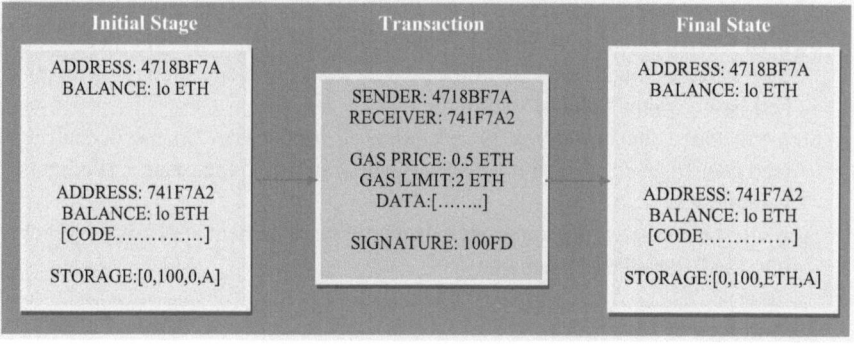

FIGURE 8.9 The Merkle tree.

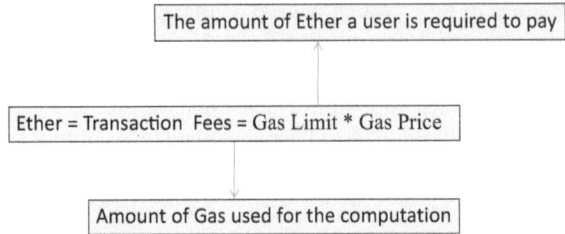

FIGURE 8.10 Ethereum logo.

Nonce: It represents the number of contracts or transactions created on an address, which also assures that each of the transactions based on the smart contract are executed only once.

Contract hash code: It is the Keccak-256 hash code of the Ethereum virtual machine.

Storage root: It is a SHA 256 hash code of the Merkle Patricia tree that carries the content of the account, which is basically the key value pair of the Ethereum account (Figure 8.13).

The transaction steps are given below:

Step i: The EOA sender initiates the transaction and signs it digitally.

Step ii: The sender sends the signed transaction to the Ethereum client.

Step iii: The next process by the client is to validate the arrived transaction, and after the validation, it's broadcasted to the decentralized peer-to-peer Ethereum network.

Step iv: The miner inserts the newly arrived transaction tin to the transaction pool.

Step v: The miner chooses a transaction from the transaction pool in a sequence and creates the new block and modifies the state of the blockchain based on the type of transaction. If the transaction is EOA, the specified value is forwarded from the sender's EOA to the receiver's EOA. If the transaction is contract account based, the contract is created and the bytecode is modified and is loaded in to the EVM.

Step vi: Unlike bitcoin, which uses PoW, Ethereum miner uses the PoS to do the verification with the help of the puzzle, and the voter's strength is identified based on the value stored in the block.

Step vii: Once the new block is created, it is broadcasted in the decentralized peer-to-peer Ethereum network by the miner, and the other participants validate it.

Step viii: Once the validation process is complete, the client appends the block into the Ethereum blockchain.

Advantages of Ethereum

The energy efficiency is high with the help of PoS.

The uptime of the network is high.

It is used to develop many decentralized applications.

Disadvantages of Ethereum
- The EVM is slow, so it creates some issues with large computation.
- As it a developing technology, it should be updated and tested based on the new bugs.
- Scalability problem remains the same as in the case of bitcoin.

8.3.3 RIPPLE PROTOCOL

8.3.3.1 What Is Ripple?

Ripple acts as both cryptocurrency and an open-source platform, which is introduced to provide a fast transaction worldwide at a cheaper cost. Ripple has its own currency, XRP, which allows any user to utilize the Ripple platform to develop his or her own network through RippleNet. Ripple uses a distributed ledger agreement with the help of validating servers and is built with the technology of blockchain [26].

8.3.3.2 Why Ripple?

It provides faster transaction with less transaction fee as compared to other cryptocurrencies. It's easily scalable and the major reason behind the introduction of Ripple is that it provides cross-border transactions with any currency, including bitcoin [26].

Features
- The transaction can be done anywhere, anytime, with any currency.
- It's more reliable as it increases visibility and reduces payment failure.
- The delay in the transaction is very less.
- The major feature is the less transaction fees even with cross-border currency.

8.3.3.3 Transaction

XRP: It is a currency used by the Ripple to transfer the value during the transaction in the Ripple network [28].

Server: The software that participates in the consensus process on the Ripple server.

Ledger: It contains the information about all the transactions and the balance amount available with all the participants of the Ripple network.

Last-closed ledger: The current status of the Ripple network is identified from this ledger and it is the most recent ledger identified in the agreement process of the current transaction.

Open ledger: This shows the current working status of all the nodes, which are initiated by the node itself. All the nodes in the Ripple network have their own open ledger.

Unique node list: All the servers in the Ripple network maintain a special list called Unique Node list; this talks about the participants of the transactions during the agreement process. The votes of the members in the Unique Node list are only considered and trusted to form the consensus.

Ripple Protocol Consensus Algorithm (RPCA)

RPCA is executed periodically by all the nodes to sustain the agreement and accuracy of the Ripple network. When the agreement is reached, the ledger currently working with it is made the last closed ledger once it is closed. This ensures that the consensus is successful and the network does not contain any fork [27,28].

The RPCA executes the following steps in each round and the steps are given next.

- Primarily, all the prior valid transactions are identified by the server before it begins with the consensus round and are made available in a public list called as "candidate set".
- The next step of the server is to merge all the server candidates set on its Unique Node list and vote on the genuineness of all the Ripple transactions.
- A minimum percentage of "Yes" votes are considered for all the transactions to pass on to the next round. The transactions that fail to get enough "Yes" are either added in the candidate list to be utilized for the next ledger or can be discarded.
- The last stage is to consider the Server's Unique Node list vote on the transaction; the expected percentage of vote is 80% and the transactions are accepted and added to the ledger once the requirement is fulfilled and the ledger is closed, which becomes the last closed ledger.

Advantages of Ripple
- It provides an open-source platform that helps the user to create his or her own blockchain in the XRPNet.
- The scalability is good as compared to bitcoin and Ethereum protocols.
- It's a cross-border cryptocurrency that enables transactions between different cryptocurrencies at very low transaction cost.

Disadvantages of Ripple
- The main disadvantage of using Ripple is that the XRP is not mined.
- The Ripple network utilizes the centralized concept in many ways.
- Even though it provides cross-border transaction, there is a lack of real-world use of this cryptocurrency.

8.3.4 LITECOIN PROTOCOL

8.3.4.1 What Is Litecoin?

It is a peer-to-peer cryptocurrency that allows for a low-cost transaction of nearly zero with less transaction time all over the world. It is an open-source decentralized payment system without any centralized authority [30].

8.3.4.2 Why Litecoin?

The block generation time is totally reduced as compared to the other cryptocurrency. The algorithm used for hashing in this Litecoin is scrypt to increase the security. The storage efficiency is also improved with a high transaction rate.

Features of Litecoin
- It is a peer-to-peer decentralized money transaction and is open to all.
- High security with the hashing method of scrypt is used in the network.
- It is an open-source platform and provides a cross border transaction with very low transaction cost and with efficient storage capabilities.

8.3.4.3 Transaction

It is similar to the other cryptocurrency, which is not introduced by government bodies; it is introduced for the money transaction of the society by the private organization. The Litcoins are generated by the process of mining, which contains the details of the processed transaction made on Litecoin. The Litecoin supply is limited unlike the other cryptocurrency. The Litecoin network creates the new blocks and the ledger contains the details of the transaction. The blocks are generated and are validated by the miner after the validation of the block is added to the blockchain.

The miners are rewarded to successfully validate the blocks, and the reward values reduce with time. When one miner is suspicious and changes the block, allowing for the same Litecoin to spend twice that is the double spending problem, it can be easily and immediately identified by the other miners in the Litecoin network.

Advantages of Litecoin
- Litecoin can handle more transactions as compared to the other cryptocurrency.
- It uses scrypt hashing method to provide a tight security.
- Provides fast transaction, which is almost nearly zero.

Disadvantage of Litecoin
- The awareness about the technology is yet to reach the society.

8.4 CONCLUSION

The initial cryptocurrency that used the blockchain technology is the bitcoin protocol. The bitcoin is the first cryptocurrency introduced, so the transition phase from the traditional system to the cryptocurrency was so difficult. When the wallet gets corrupted or lost, the entire currency will be donated to the system. This is the major drawback of the cryptocurrency. The Ethereum protocol is the next popular cryptocurrency that brings the advantage as an open-source platform to create more decentralized applications. The next popular cryptocurrency with the open-source platform is the Ripple protocol. This protocol concentrated on the efficient utilization of energy and fast transaction with the cross-border transaction. The last protocol discussed is the Litecoin, which also performs cross-border transaction, with the transaction cost being nearly zero; it is also an open-source platform that provides a high security and performs transactions faster as compared to the cryptocurrencies.

REFERENCES

1. Zibin Zheng, Shaoan Xie, "Blockchain challenges and opportunities: A survey," *International Journal of Web and Grid Services*, vol. 14, no. 4, 2, October 2018.

2. Francisca Adoma Acheampong, "Big data, machine learning and the blockchain technology: An overview," *International Journal of Computer Applications*, vol. 180, no. 20, 0975–8887, March 2018.

3. X. Li, et al., "A survey on the security of blockchain systems," *Future Generation Computer Systems*, 2017, doi: 10.1016/j.future.2017.08.020.

4. Shubhani Aggarwal "Blockchain for smart communities: Applications, challenges and opportunities," *Journal of Network and Computer Applications*, vol. 144, 13–48, 2019. ISSN: 1084-8045.

5. Remya Stephen, Aneena Alex, "A review on blockchain security," IOP Conference Series: Materials Science and Engineering, vol. 396, 012030, 2018.

6. Shanhong Liu, Size of the Bitcoin blockchain from 2010 to 2019, by quarter (in megabytes), https://www.statista.com/statistics/647523/worldwide-bitcoin-blockchain-size/

7. Back, Hashcash—a denial of service counter-measure, 2002. [Online], available at: http://www.hashcash.org/papers/hashcash.pdf

8. Mauro Conti, Sandeep Kumar E., Chhagan Lal, Sushmita Ruj, "A survey on security and privacy issues of bitcoin," *IEEE Communications Surveys &Tutorials*, vol. 20, no. 4, Fourth Quarter, 2018.

9. Satoshi Nakamoto, "Bitcoin: A peer-to-peer electronic cash system," 2008, available at: https://bitcoin.org/bitcoin.pdf

10. Dejan Vujičić, Dijana Jagodić, Siniša Ranđić, "Blockchain technology, bitcoin, and Ethereum: A brief overview," *17th International Symposium INFOTEH-JAHORINA*, 21–23 March 2018.

11. Florian Tschorsch, Björn Scheuermann, "Bitcoin and beyond: A technical survey on decentralized digital currencies," *IEEE Communications Surveys & Tutorials*, vol. 18, no. 3, 2084–2123, March 2016.

12. Ethereum Community, "A next-generation smart contract and decentralized application platform," *White Paper*, available at: https://github.com/ethereum/wiki/wiki/White-Paper

13. Ralph C Merkle, "A digital signature based on a conventional encryption function," In: Pomerance C. (eds) *Advances in Cryptology — CRYPTO '87. CRYPTO 1987.* Lecture Notes in Computer Science, vol. 293. Springer, Berlin, Heidelberg, pp. 369–378, 1987.

14. Ralph C Merkle, "Protocols for public key cryptosystems," in *Proc. 1980 Symposium on Security and Privacy, IEEE Computer Society*, pp. 122–133, April 1980.

15. Transaction malleability, available at: https://en.bitcoin.it/wiki/Transaction_malleability

16. Cointelegraph, SegWit explained, available at: https://cointelegraph.com/explained/segwit-explained

17. Alyssa Hertig, "SegWit goes live: why Bitcoin's big upgrade is a blockchain game-changer," August 2017, available at: https://www.coindesk.com/50-blocks-segwit-bitcoins-coming-upgradeblockchain-game-changer/

18. Sujha Sundararajan, "Blockstream launches micropayments processing system for Bitcoin apps," January 2018, available at: https://www.coindesk.com/blockstream-launches-micropaymentsprocessing-system-for-bitcoin-apps/

19. Archana M. Naware, " Bitcoins, its advantages and security threats," *International Journal of Advanced Research in Computer Engineering & Technology (IJARCET)*, vol. 5, no. 6, 11, June 2016.

20. Huashan Chen, Marcus Pendleton, Laurent Njilla, Shouhuai Xu, "A survey on ethereum systems security: Vulnerabilities, attacks and defenses," arXiv:1908.04507v1 [cs.CR], August 2019.

21. Ingo Weber, Vincent Gramoli, Alexander Ponomarev, Mark Staples, Ralph Holz, An Binh Tran, Paul Rimba, On Availability for Blockchain-Based Systems. 64-73. (2017) 10.1109/SRDS.2017.15.

22. Chris Dannen, Introducing Ethereum and Solidity: Foundations of Cryptocurrency and Blockchain Programming for Biginners. ISBN-13 (pbk): 978-1-4842-2534-9, ISBN-13 (electronic): 978-1-4842-2535-6, doi: 10.1007/978-1-4842-2535-6.

23. Nicola Atzei, Massimo Bartoletti, Tiziana Cimoli, "A survey of attacks on Ethereum smart contracts (SoK)," *Conference on Principles of Security and Trust.* POST 2017. Lecture Notes in Computer Science, vol. 10204. Springer, Berlin,

24. Gavin Wood, "Ethereum: A secure decentralised generalised transaction ledger, Byzantium version," 2018, available at: https://ethereum.github.io/yellowpaper/paper.pdf

25. Yonatan Sompolinsky, Aviv Zohar, "Secure high-rate transaction processing in Bitcoin," Financial Cryptography, 507–527, 2015.

26. Chris Larsen, the content is available at https://ripple.com/xrp/

27. Peter Todd, "Ripple protocol consensus algorithm review," May 11th 2015, available at: https://raw.githubusercontent.com/petertodd/ripple-consensus-analysis-paper/master/paper.pdf

28. David Schwartz, Noah Youngs, Arthur Britto, "The ripple protocol consensus algorithm," available at: https://ripple.com/files/ripple_consensus_whitepaper.pdf

29. Charlie Lee, the content is available at https://litecoin.org/

30. https://www.investopedia.com/articles/investing/040515/what-litecoin-and-how-does-it-work.asp

9 FGUGChain
A Blockchain Application Framework with Secure Computation

A. Anasuya Threse Innocent and G. Prakash
Amrita School of Engineering, Amrita Vishwa Vidyapeetham

CONTENTS

9.1 INTRODUCTION

9.1.1 BLOCKCHAIN

Blockchain is a decentralized computation and information sharing platform that enables multiple authoritative domains that do not trust each other to cooperate, coordinate, and collaborate in a rational decision-making process. In a typical blockchain architecture, every node (the system/computer of each party) maintains a local copy of the global datasheet. The system ensures consistency among the local copies by ensuring that (i) the local copies at every node are identical and (ii) the local copies are always based on the global information.

The local information stored on each party's computer is called a public ledger, which can be considered as a database available with everyone, containing the historical data that can be used for future computations. The public ledger stores all the global transactions that are synchronized. For example, consider four parties named Anna, Ben, Dav, and Emma who are involved in distributed banking. They have Rs. 5,000 each in their accounts, and all four have a public ledger stored at their local systems. Now, consider the following transactions:

1. Anna sends Rs. 2,000 to Ben – accepted and updated in everyone's ledger
2. Ben sends Rs. 4,000 to Dav – accepted and updated in everyone's ledger
3. Anna sends Rs. 4,000 to Emma – transaction rejected by remaining three and is not successful

The sample ledgers of Anna, Ben, Dav, and Emma after the above transactions are shown in Figure 9.1.

Here, the first two transactions are successful; they were verified and accepted by all the parties. But the third transaction, *Anna → Emma: Rs. 4,000* was not successful. When Anna broadcasts her request, other parties check the balance in her account by checking transactions committed in their public ledger. From that they learn that Anna doesn't have the required balance for the transaction. Hence, all the other parties reject the third transaction mentioned above, and so this transaction is not recorded in their public ledger.

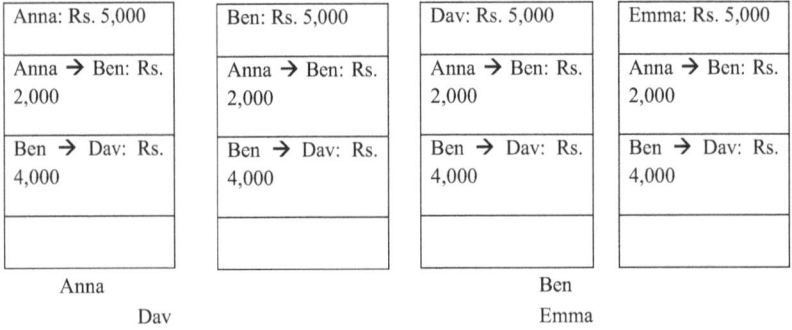

FIGURE 9.1 Sample public ledgers of the parties involved.

The basic requirements for blockchain are as follows:

- **Protocols for commitment**

  There should be a proper mechanism to ensure that every valid transaction from the clients is committed and included in the blockchain within a finite time. For example, the first two transactions mentioned in the above example have to be verified one after the other and committed before the next transaction starts.
- **Consensus**

  There should be a mechanism to ensure that the local copies of the public ledger are consistent and updated.
- **Security**

  When one party broadcasts the information, other parties involved should be able to validate that the information is tamperproof. If a client acts maliciously or if a client is compromised, it should be identified by the other parties immediately.
- **Privacy and authentication**

  As a number of parties are involved in a blockchain, authenticity and privacy of the data should be maintained.

According to Iansiti and Lakhani (2017), blockchain can be defined as "an open distributed/decentralized ledger that can record transactions between the parties efficiently and in a verifiable and permanent way". The public ledger used is accessible to all and is not controlled by a single party. The transactions between the parties should be fast and scalable, and everyone involved can check the validity of information. Also, the information is persistent. They have also briefed the working of blockchain. The five basic principles of blockchain technology mentioned by Iansiti and Lakhani are extracted from Iansiti and Lakhani (2017) and is shown in Figure 9.2.

The cryptographic primitive used to build the blockchain is a hash function. Every block stores the hash value of the previous block and thus the hash function acts as a link to form the chain of blocks, resulting in a blockchain. This idea of using hash values to form a chain of blocks is derived from Harber and Stometta (1990). The Merkle hash trees by Merkle, developed in 1979 (Merkle, 1982), is combined with the work of Harber and Stometta (1990) to develop efficient time stamping of several documents to form one block (Bayer et al., 1993). This invention led to the discovery of bitcoin by Nakamoto (2008), which is built on the foundation of blockchain technology.

Bitcoin is a completely decentralized peer-to-peer permissionless cryptocurrency. As it is permissionless, anyone without any identity or login credentials can use it anywhere, without any access control. This leads to legal issues in many counties, but the technology behind it, the blockchain, can be used for a number of useful applications. Blockchain technology can be used not only in the financial sector but also in almost all the distributed, decentralized applications. Governments, industries, academia, and agriculture domain are moving toward blockchain, and this revolution is called Blockchain 2.0.

1. **Distributed Database/Public Ledger**

   Each party involved in the computation of blockchain has access to the entire database/public ledger and its complete history. No single party controls the data or the information. Every party can verify the records of its transaction partners directly, without an inerymediary or any central controller.

2. **Peer-to-Peer Transmission**

   Communication between the peers takes place directly, instead of using a central node. Every node/ party involved have the capacity to store and forward the information to all other parties.

3. **Transparency with Pseudonymity**

   Anyone who has access to the system can view the transactions and the values associated them. A unique 30+ character address is assigned to each party involved in blockchain, which is used for transactions. Users/parties can either remain anonymous or provide proof of their identy to others.

4. **Irreversibility of Records**

   The records cannot be altered after the transaction is verified and entered in the database. All the transactions are linked and updated in the blocks of the blockchain. Multiple algorithms and approaches are used to ensure permanent recording in database in a chronological order, available to all others on the network.

5. **Computational Logic**

   The public ledger is maintained in digital form. This makes the transactions linked to computational logic, which can be programmed. Algorithms and rules can be programmed in such a way that transactions between the nodes are automatically triggered.

FIGURE 9.2   Five basic principles of blockchain (Iansiti and Lakhani, 2017).

The Blockchain 2.0 involves smart contracts to achieve decentralization without a middleman. Smart contract introduced by Szabo in the late 1990s (Szabo, 1997) is an automated computerized protocol used for digitally facilitating, verifying, or enforcing the negotiation on performance of a legal contract by avoiding intermediates and directly validating the contract over a decentralized platform in a faster, cheaper, and more secure manner. Smart contracts are:

i. **Immutable**: It means that no party will be able to change the contract once it is fixed and written to the public ledger – blockchain.
ii. **Distributed**: It means that there will be no middleman, and all the steps of the contract can be validated by every participating party. Hence, no one can claim later that the contract was not validated.

These properties of smart contract make it suitable to be used in blockchain, as the blocks in it are immutable and the information is open to all the parties involved; anyone can check and validate the information.

The original blockchain design, which is permissionless, is mainly suitable for financial applications and is tamperproof. Here, tamperproof means that it is extremely difficult to make changes in the blockchain, and the difficulty increases proportionally with the chain length. Bitcoin is the best example for permissionless blockchain. Each block in bitcoin consists of an address/header and data/information. Address in bitcoin is synonymous to the "Account" in a bank. With this address, authenticity is ensured, that is, only the intended party can receive and decrypt the transaction sent by the respective party and accept it. It also provides privacy to some extent, that is, information is encrypted when transferring between the nodes; and also pseudonyms are used, but the information is exposed on the nodes for verification; hence making them unsuitable for working with private data.

Multiple cryptocurrencies have been developed using blockchain after the invention of bitcoin (Nakamoto, 2008). Some of them are listed here: Litecoin (Lee, 2011), Peercoin (King and Nadal, 2012), Emercoin (2013), Gridcoin (Halford, 2014), Stellar (Mazieres, 2015), Ethereum (Buterin, 2016), Omni (Willett et al., 2016), Bitcoin Cash (2017), Waves (Anderson et al., 2017), EOS (Grigg , 2017), Cardano (Hoskinson, 2017), USD Coin (2018), and the list keeps growing. Of the cryptocurrencies developed, some act just as cryptocurrency, while others act as a platform to develop blockchain applications.

Researchers are constantly working on improving the efficiency of blockchain by improving the throughput and scalability, and also working on adding privacy in order to use it for a wide range of applications. As an attempt to include privacy, permissioned/private blockchains are introduced, sacrificing the decentralized nature of blockchain. Eris, Ripple, and Hyperledger are a few to name. The private blockchain uses an access control layer to allow only the permissioned parties to involve in transactions. Even though an extra layer of privacy is ensured in private blockchains by the help of private transactions, it moves toward a central entity as verification of transactions and controls are in the hands of the owner of the particular private blockchain.

To achieve privacy of data in blockchain, preserving its decentralization nature, another major breakthrough in theoretical computer science, namely, *Secure Computation*, is used along with blockchain.

9.1.2 SECURE COMPUTATION

Secure computation, the term introduced by Yao in early 1980s (Yao, 1982), is a part of cryptography in which two or more parties with private inputs wish to compute some joint function of their inputs. The readers should not get confused about secure computation with secure communication. In secure communication, data from one party are securely communicated to the other party through a channel with the help of encryption/decryption algorithms. The private data from one party (sender) is encrypted and sent to the intended party – the receiver who can decrypt and access the private data of the sender. But in secure computation, private inputs from the parties involved are used to compute a function without exposing the private data. A party involved in secure computation will know only the output of the function and his or her private data. In a nutshell, the problem behind secure computation can be stated as follows:

> Consider a set of parties who do not trust each other, nor the channels by which they communicate. Still, the parties wish to correctly compute some common function of their local inputs while keeping their local data as private as possible.

In other words, "combining information while protecting it as much as possible is termed as secure computation". When the number of parties involved in computation are only two, then it is called as *Secure Two-Party Computation* or simply as *Two-Party Computation*, denoted as 2PC.

When the number of parties involved are three or more, to compute a common functionality without revealing their private data, it is termed as *Secure Multiparty Computation,* or simply as *Multiparty Computation*, denoted by MPC. Election can be considered as an example of multiparty computation; voters want that their votes (and their identity) be kept secret, and at the same time it has to be counted. Another example is conducting a study of a new disease without revealing the individual test reports of patients or their identities. The baseline is, computation has to be done on private data without exposing them, and the outcome should reveal only the desired output, and whatever minimal information leaked by it. A number of day-to-day life problems ranging from as simple as coin tossing and mutual agreement to complex applications such as e-voting, electronic auctions, private data retrieval, analysis of sensitive information to conduct research on it without exposing them, privacy preserving biometric identification, private editing in cloud, etc. can be solved by secure computation.

Secure computation protocols are classified into four dimensions depending on the circuit model, network model, distrust model, and adversary model (Patra, 2015). Regardless of the dimension, the first step in any secure computation protocol design is the construction of *Garbled Circuit* (GC). Since the mathematical expression underlying any distributed computing problem can be represented either as a Boolean circuit or as an arithmetic circuit over a finite field, the choice of circuit model depends on the function to be computed. Nonlinear operations such as comparison, greater than, less than, etc. are represented in Boolean circuits.

Garbled circuit construction on computational setting with underlying Boolean circuits with two-input one-output gates are described in Lindell and Pinkas (2009). In this classical circuit construction, two random value bit-strings corresponding to 0 bit and 1 bit inputs are assigned to each of the input wires. These random values/ keys are encrypted according to the gate type. Per wire, encryption is carried out twice; and on the values obtained, simple permutation is applied to form the garbled table entries, resulting in garbled circuit of the required functionality (Innocent and Sangeeta, 2014a; 2014b). Yao's garbled circuit construction for a single gate is shown in Appendix 9.1.

Yao's design uses double encryption per wire (Lindell and Pinkas, 2009), which increases the computation time. As the size of circuit increases, complexity also increases, resulting in impractical protocols. To overcome this bottleneck, improvement in efficiency is needed, which clearly indicates the need for optimization in garbled circuit design. The following are well-known major optimizations on garbled circuit design.

9.1.2.1 Point-and-Permute Technique

After the garbled circuit is generated and sent to the evaluator, evaluation takes place in order to find the required output. The correct row to be decrypted has to be identified with some mechanism in order to maintain correctness and avoid unnecessary decryption of all the rows during evaluation. It is achieved with the help of point-and-permute technique (Naor et al., 1999), which provides a pointer to the row to be evaluated. The color bit added during garbling of gates provides a pointer to the row, which has to be decrypted during evaluation, to maintain correctness.

9.1.2.2 Free-XOR Technique

One of the best optimizations in garble circuit construction with Boolean circuits is the free-XOR technique (Kolesnikov and Schneider, 2008). This skips the complex encryption/decryption calls whenever a XOR gate is encountered. The output of XOR gate itself is scrambled, i.e., does $\oplus$ operation on the inputs, which is a basic encryption; hence, complex encryptions are avoided on XOR gates. The working is as follows: three or more input XOR gates are converted into two input XOR gates, and free-XOR operation is carried out. Considering X_a^0, X_b^0 as the input wire values of XOR gate corresponding to the inputs (0, 0), the output is obtained as $X_{out}^0 = X_a^0 \oplus X_b^0$, $X_{out}^1 = X_{out}^0 \oplus R$, where R is a globally used random value on protocol. They proved simulation-based security of their protocol and showed that the simulator output is indistinguishable to the original protocol output.

9.1.2.3 Garbled Row Reduction

According to the Garbled Row Reduction technique (GRR) (Pinkas et al., 2009), the size of GC is substantially reduced concerning the number of table entries. In this technique, one garbled value is defined as a function of two input wire values. It just replaces the consideration of the two garbled values defined randomly as part of the output wires. The garbled table of the gate therefore need not store the value corresponding to this, and during evaluation phase, without consulting the garbled table the corresponding garbled output can be computed. Even though they claim that GRR and free-XOR are noncompatible, they are able to show improvement in efficiency, as one-fourth of the message to be communicated is reduced, which results in improved communication complexity as well a marginal value in computation complexity.

9.1.2.4 Topological Sorting of Gates

Might Be Evil project (Huang et al., 2011) showed how topological sorting of gates can be used and circuit generation can be pipelined without storing the entire garbled circuit. It also introduced oblivious transfer extension along with the usage of point-and-permute, garbled row reduction, and free-XOR optimizations.

9.1.2.5 Dual-Key Cipher

Bellare et al. (2012) changed the view of GC as a tool for secure computation protocol design to a complete cryptographic goal. Dual-key cipher (DKC) introduced by them for faster construction of garbled circuits using Boolean gates makes only one complex encryption call per wire, and thus reduces the computation time drastically. They call the garbled circuit construction garbling scheme, and as a separate cryptographic goal it needs optimization. They prove indistinguishability-based notion of security and simulation-based security for the garbling scheme and also show that simulation-based security always implies indistinguishability-based security.

9.1.2.6 Simple Circuit Description (SCD), AES-NI

The JustGarble system developed by Bellare et al. (2013) based on DKC cipher with fixed-key AES makes a single call to a fixed permutation – for gate evaluation. It also introduces a simple representation of garbled circuit, named as Simple Circuit Description (SCD), in which the gates are represented as an array. Here, as the gates are represented just as arrays arranged in a topological order, the need for communication between gates as objects are scraped, making the system lightweight, more efficient, and cost-effective. The use of embedded AES makes new instructions speed up the process. JustGarble enquires only one AES call per gate for evaluation. It stands as a benchmark for further optimizations on garbled circuit construction using Boolean gates and acts as the base for comparison of new optimizations/frameworks.

9.1.2.7 Other Optimizations

Some other known optimizations/techniques to improve bit and time complexities are as follows:

- **FleXOR** (Kolesnikov et al., 2014)

The flexible garbling of XOR gates allows to have 0–2 ciphertexts for the XOR gate. It also combines the AND gate optimizations with free-XOR.

- **Half Gates** (Zahur et al., 2015)

It is designed for privacy-free setting. It produces only two ciphertexts per AND gate and is compatible with the free-XOR technique.

- **Garbling Gadgets** (Ball et al., 2016)

It is designed to work on Boolean and arithmetic circuits. When used for Boolean circuits, it generates $n + 1$ ciphertexts instead of $2n$ ciphertexts for the corresponding n inputs. Ball et al. have also theoretically claimed that a single AND gate can be garbled to a single ciphertext.

- **Reusable Garbled Circuit**

Reusability of garbled circuit construction with input and circuit privacy is proposed by Wang (2016). They have also proposed a reusable garbled circuit with input privacy and circuit privacy, and have proved the security using indistinguishability-based security scheme. In 2017, they have introduced reusable garbled circuit construction with fully homomorphic encryption (Wang et al., 2017) and claims of constant communication complexity.

- **Size-Zero Garbled Circuits**

Kondi and Patra (2017) proposed garbling of Boolean circuits without producing ciphertext in privacy-free setting, resulting in size-zero garbled circuits.

- **Authenticated Garbling**

An improvement in efficiency on malicious adversary setting by a factor of 1.5 and 2 on communication and computation complexity is shown to be achieved by the use of leaky AND triples by Katz et al. (2018). They have showed an application with authenticated garbling on WRK protocol (Wang et al., 2017), by avoiding MAC on each garbled row, and which is also compatible with half gates.

Most of the above said optimizations under the "other optimizations" subsection haven't yet found major implementation outbreak on secure computation practical frameworks. Another two optimizations used in the FGUGChain are Batch-Key Cipher (BKC) and the universal gate optimization (Innocent et al., 2019).

9.1.2.8 Batch-Key Cipher (BKC)

Batch-key cipher with fixed-key encryption for garbled circuit construction makes complex encryption calls once per block of gates and achieves 36% improvement in efficiency in terms of execution time of protocol, without compromising the security requirements. Universal gates on garbled circuit construction (Innocent et al., 2019a) are built on the FastGarble framework (Innocent et al.), which is explained in the next section. Both BKC and universal gates together have been proved to be secure and they speedup the garbling time by 66%.

Researchers are constantly working on improving time complexity and communication complexity of secure computation protocols as any distributed decentralized application can be implemented with secure computation. Even though secure computation can be considered as the best solution for privacy of sensitive data, monitoring the parties involved in a computation is a tedious task, as they have to be constantly checked for adversaries, and only the honest parties have to be successful in computation of the required functionality. To ensure honesty of the parties involved, blockchain concept is the best choice, and hence combining both the technologies will be a win–win situation for both. Blockchain will ensure that only the honest parties are part of computation, and secure computation ensures privacy of sensitive data. The next section gives a short description of research on blockchain with secure computation to achieve blockchain with privacy.

9.1.3 RESEARCH ON BLOCKCHAIN WITH SECURE COMPUTATION

In 2015, the Enigma platform (Zyskind et al., 2015) made its first attempt to combine secure computation with blockchain technology to introduce privacy in blockchain. The off-chain or off-line phase in it does the complex multiparty computation, and the blockchain operations are carried out in the on-chain or the online phase. In the previous year, Andrychowicz et al. (2014) showed how bitcoin can be used in the area of secure computation, and also Bentov and Kumaresan (2014) showed how to use bitcoin to design fair secure computation protocols.

CoinParty (Ziegeldorf et al., 2015) provides a mixing service for bitcoin with multiparty computation. It allows to distributedly create bitcoin addresses for fund transfers using threshold transactions – meaning that only when a majority of parties involved agree to proceed with the transaction. The Hawk compiler (Kosba et al., 2016) introduced privacy preserving smart contracts with the help of cryptographic primitives and also confirmed that multiparty computation-based approach can produce fair transactions. Inventors of Hawk have clearly mentioned that "blockchain can be trusted for correctness and availability, but not trusted for privacy", and explained the need for secure computation to achieve privacy in blockchain.

Noyes (2016) showed a peer-to-peer multiparty computation implementation on the secret chat/social networking application Pandora, and also compared the performance with secure computation frameworks FairPlayMP (Ben-David et al., 2008) and VIFF (Geisler, 2007). AntNest (Zhou et al., 2018) provides a multiparty computation framework with verifiability of parties involved with blockchain. AntNest uses fully homomorphic noninteractive verifiable secret sharing for multiparty computation part and showed improvement in execution time of applications. Another framework Raziel (Sanchez, 2018) used multiparty computation using zero-knowledge proof to provide privacy, correctness, and verifiability for smart contracts on blockchain. Raziel also gives practical implementation of applications such as crowdfunding and double auctions. PlatON (2019) aims to develop an efficient full-stack decentralized infrastructure with privacy, correctness, verifiability, and unforgetability. Wanchain (2019) aims to build a distributed bank by providing a cross-chain platform with privacy of multiparty computation.

9.2 FGUGCHAIN FRAMEWORK – A BLOCKCHAIN WITH PRIVACY

The FGUGChain framework introduced by Innocent and Prakash (2019) is described in detail in this section. FGUGChain (*FastGarble Universal Gates Chain*) comprises of online phase and off-line phase to achieve blockchain with privacy and faster computation using an efficient garbled circuit construction. The FGUGChain architecture is shown in Figure 9.3, and the components are explained in the following sections.

9.2.1 ONLINE PHASE

The online phase or the on-chain consensus of FGUGChain does the lightweight blockchain operations. The process is carried out in such a manner that the original data are never exposed to the parties involved in the transaction, as well data is not stored on nodes. The major components in online phase are smart contract metadata, computation function of the application, and the blockchain nodes.

9.2.1.1 Smart Contract Metadata

Smart contract is the set of promises or the business logic between the parties involved in transactions. Smart contract is a software that runs on the public ledger to encode assets and has transaction instructions for modifying assets. The contract metadata

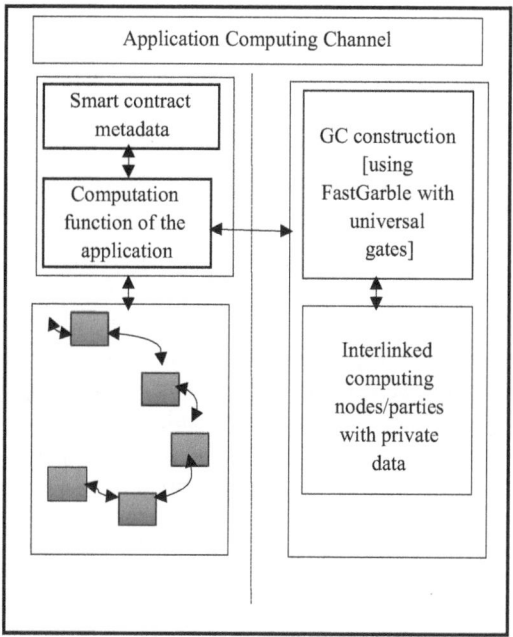

FIGURE 9.3 FGUGChain framework (Innocent and Prakash, 2019).

contains the information about the contract, which is stored on a location, and its address is kept on the contract. Whoever has access to the address can access the metadata and get the information from it. The smart contract used in FGUGChain is stateless and similar to that of PlatON. It does not store any states of the chain, or reveals data to the computing nodes, i.e., the parties involved. When smart contract metadata is executed, the input data come from the local databases of the off-chain nodes participating in multiparty computation. The actual computation is broken down into multiple subtasks and are distributed to the computing nodes without revealing the data. Ownership of the parties involved and privacy are guaranteed by the multiparty computation protocol used.

9.2.1.2 Computation Function of the Application

Any application that can be implemented can be represented as a mathematical function, which is called the computation function of the application, and any mathematical function can be represented in Boolean circuits. Computation function is defined by the smart contract metadata along with the input and output parameters. This computation function is converted into a Boolean circuit, divided into multiple sub circuits and is sent for garbled circuit construction in off-line phase, and the garbled circuits generated are linked to it.

9.2.1.3 Blockchain

The computation function defined by the smart contract metadata is split and distributed to multiple computing nodes/parties by the block producers while generating the blocks. The computing nodes return the result as well as the proof of computation, and they are also packed into the blocks by the producer. Validity of the block is determined by other nodes, just by verifying the proof, which reduces the complex, redundant computations on each node, and hence the block verification time. A new block is added to the chain once the verification process is completed.

9.2.2 OFF-LINE PHASE

The off-line phase or the off-chain computations carry out the complex multiparty computation protocol execution. Yao's GC construction (Lindel and Pinkas, 2009; Innocent and Sangeeta, 2014a; 2014b) plays a vital role in multiparty computation protocols. The underlying computation function is implemented into a Boolean circuit, divided into multiple subcircuits, and are then sent to computing nodes and are executed in parallel. Each subcircuit is sent to multiple computing nodes to maintain certain level of redundancy and availability guarantee. The garbled circuit constructor selects random tokens for each wire in the Boolean circuit, encrypts the output token with that of input tokens for each gate, the encrypted tokens are garbled per gate, and thus the garbled circuit is generated (Lindell and Pinkas, 2009; Innocent and Sangeeta, 2014a, 2014b; Bellare et al., 2012). The garbled circuit evaluator uses oblivious transfer protocol and deciphers the garbled circuit to obtain the required result (Lindell and Pinkas, 2009). The components of off-line phase are described in the following sections.

9.2.2.1 Garbled Circuit Construction Using FastGarble with Universal Gates

The garbled circuit construction here uses the FastGarble with universal gates framework (Innocent et al., 2019a). Here, instead of different types of gates, any one of the universal gates (NAND/NOR) is used for Boolean circuit generation, and hence for garbled circuit construction. Also the batch-key cipher optimization for garbled circuit construction is used along with point-and-permute (Noar et al., 1999), garbled-row reduction (Pinkas et al., 2009), dual-key cipher with fixed-key block cipher with Advanced Encryption Standards-New Instructions (AES-NI) and Simple Circuit Description (SCD) (Bellare et al., 2013) optimizations. The FastGarble with universal gates framework is proved to be $\cong 66\%$ more efficient in terms of time complexity measured by execution time of protocol, and also requires only two ciphertexts per gate for communications (Innocent et al., 2019a). Overall, the use of FastGarble with universal gates framework improves the efficiency of multiparty computation application.

9.2.2.1.1 Universal Gates on Garbled Circuit Construction

The NOR and NAND logic gates are called the universal gates as the other logic gates, namely, AND, OR, XOR, and NOT can be derived from these gates. FastGarble with universal gates framework is built on FastGarble framework and uses only the universal gates for Boolean circuit construction. Circuit construction with any one universal gate comes with the advantage that a single type of gate is used for the circuit construction. This eliminates the need for gate array G in SCD. FastGarble with universal gates framework is built on FastGarble framework and uses only the universal gates for Boolean circuit construction. As an example, AND circuit suitable for SCD representation built only with NOR gates is shown in Appendix 9.2, and an illustrated example is given by Innocent et al. (2019a). The gate-type array G in the SCD format of JustGarble (Bellare et al., 2013) $f_{JG} = (n,m,q,A,B,\underline{G})$ can be removed to form the SCD circuit representation of FastGarble with universal gates having 5-tuples as $f = (n,m,q,A,B)$. The storage size required for the application reduces as the gate array need not be stored; instead, a constant value corresponding to the gate type is used in protocol. Also, checking of respective gate type is not needed for every gate during garbling, which contributes marginally for the improvement of time complexity. FastGarble with universal gates uses the point-and-permute, garbled row reduction of two rows per gate (GRR-2), SCD, AES-NI, and BKC. Free-XOR technique is neglected as there are no XOR gates available in the circuit. As batch-key cipher is used, it requires only two complex encryption/decryption calls per batch of gates. Use of the single gate type (either NAND or NOR) always ensures the encryption of every first gate of the batch with dual-key cipher, and thus new encryption per batch is applied. Also, even though representing the Boolean circuit with universal gates increases the number of gates, usage of batch-key cipher and GRR-2 design makes it approximately 66% faster than the available protocols for garbled circuit construction. This makes it a better choice to be used in FGUGChain framework.

9.2.2.2   Interlinked Computing Nodes/Parties with Private Data

Interlinked computing nodes are the parties involved in multiparty computation with their private data. With this FGUGChain framework, they are able to compute on their private data without revealing it to even a trusted third party on a decentralized environment.

9.3   CONCLUSION

The FGUGChain framework discussed in this chapter gives a theoretical exposure to the framework, and it can be used to develop various distributed decentralized applications with privacy. The FastGarble with universal gates framework incorporated to include privacy by the use of secure computation is 66% faster than the existing practical implementations on secure computation. As well, it uses only two ciphertexts per gate and is proven to be secure. The blockchain technology provides better decentralization and allows only the honest parties to be part of the computing environment. Thus, we can say that FGUGChain acts as a platform for generating distributed decentralized applications with privacy by intertwining secure computation with blockchain technology.

APPENDIX 9.1

Yao's Garbled Circuit Construction

The first step in Yao's 2PC design under computational model of security is the construction of GC, followed by OT. While using Boolean circuits, the underlying computation function $f(x,y)$ is represented in circuit form by the first party Alice.

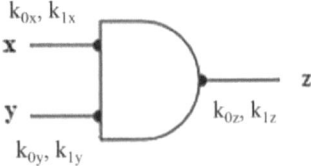

FIGURE 9.4 The inputs and output along with the key labels assigned by the first party Alice for a 2-input 1-output AND gate.

She generates the garbled circuit and sends that circuit to the second party Bob. Figure 9.4 shows a 2-input 1-output AND gate, to which the garbling steps (Lindell and Pinkas, 2009) are explained as follows:

Step 1: For each wire of the gate, Alice chooses two random keys with length corresponding to the size of encryption algorithm used.
Step 2: Alice encrypts the truth table and garbles the encrypted values.
Step 3: Alice sends garbled truth table to the other party, Bob.

Communication between Alice and Bob are following the Steps 1–3. In Step 1, Alice randomly selects two keys per wires, say (k_{0x}, k_{1x}), (k_{0y}, k_{1y}) corresponding to the inputs x, y and (k_{0z}, k_{1z}) corresponding to the output wire z (refer to Figure 9.4). Truth table of AND gate is shown in Table 9.1, and Table 9.2 shows the encrypted value, which will be garbled and the value obtained will be sent to Bob.

TABLE 9.1
The Truth Table of AND Gate

Inputs		Output
x	Y	Z
0	0	0
0	1	0
1	0	0
1	1	1

TABLE 9.2
Double Encryption on AND Gate

Wire 1 Input, x	Wire 2 Input, y	Wire 3 Output, z	Encrypted Value
k_{0x}	k_{0y}	k_{0z}	$E_{k_{0x}}\left(E_{k_{0y}}\left(k_{0z}\right)\right)$
k_{0x}	k_{1y}	k_{0z}	$E_{k_{0x}}\left(E_{k_{1y}}\left(k_{0z}\right)\right)$
k_{1x}	k_{0y}	k_{0z}	$E_{k_{1x}}\left(E_{k_{0y}}\left(k_{0z}\right)\right)$
k_{1x}	k_{1y}	k_{1z}	$E_{k_{1x}}\left(E_{k_{1y}}\left(k_{1z}\right)\right)$

Here, the encryption carried out is double encryption, i.e., per wire complex encryption call is made twice. This creates a bottleneck on efficiency and makes the protocol impractical. The step-by-step working of GC construction and evaluation is shown with illustrated example by Innocent and Sangeeta (2014a; 2014b).

APPENDIX 9.2

AND Circuit in SCD

For an example to show the SCD circuit construction using NOR gates, AND circuit is used, which is nothing but the simple AND gate function giving output 1 when both A and B are 1. Figure 9.5 shows the AND circuit designed only with NOR gates, and is suitable for representation in SCD format. The truth table for the AND circuit is shown in Table 9.3.

SCD format of any function can be represented using a 7-tuple, $f'_{old} = (n, m, q, A, B, G, O)$. Name and meaning of parameters in it and their respected values for AND circuit shown in Figure 9.5 are given next:

Number of inputs, $n = 2$
Number of outputs, $m = 1$
Number of gates, $q = 3$
First input wires to the gates, $A[] = \{1, 2, 3\}$
Second input wires to the gates, $B[] = \{1, 2, 4\}$
Gate type, $G = 1$
Output wire, $O[] = \{5\}$

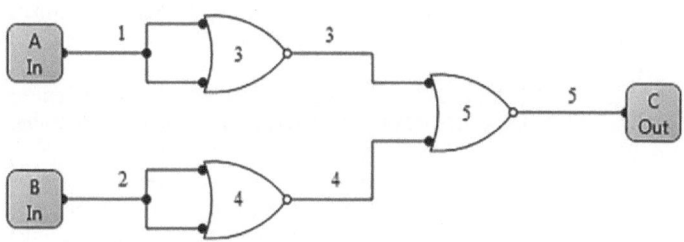

FIGURE 9.5 AND circuit generated only with NOR gates, and suitable for SCD format.

TABLE 9.3
Truth Table of AND Circuit

Inputs		Output
A	B	C
0	0	0
0	1	0
1	0	0
1	1	1

The above values can be included to get SCD format of circuit representation as in Equation 9.1.

$$f'_{old} = \left(2,\ 1,\ 5,\ [1,\ 2,\ 3],\ [1,\ 2,\ 4],\ 1,[5]\right) \tag{9.1}$$

The gate type is found using the formula in Equation 9.2.

$$b_{11} \times 8 + b_{10} \times 4 + b_{01} \times 2 + b_{00} \times 1,\ b_{ij} = f(i,j) \tag{9.2}$$

For example, NOR gate $\rightarrow 0 \times 8 + 0 \times 4 + 0 \times 2 + 1 \times 1 = 1$. Here, b_{ij} represents the corresponding truth table values of the particular gates with the given values of i and j. The gate-type array G is replaced by a constant value (here, $G = 1$) as a single type of gate is used, which can be directly initialized on the program. By removing G from the SCD circuit representation, the updated SCD representation for FastGarble with universal gates used in this chapter is of 6-tuple $f' = (n,m,q,A,B,O)$, as shown in Equation 9.3, with initial circuit as a 5-tuple $f = (n,m,q,A,B)$.

$$f' = \left(2,\ 1,\ 5,\ [1,\ 2,\ 3],\ [1,\ 2,\ 4],\ [5]\right) \tag{9.3}$$

REFERENCES

Andersen MP, Kolb J, Chen K, Fierro G, Culler DE, Popa RA. Wave: A decentralized authorization system for IoT via blockchain smart contracts. *EECS Department, University of California, Berkeley, Tech. Rep.* UCB/EECS-2017–234. 2017 Dec 29.

Andrychowicz M, Dziembowski S, Malinowski D, Mazurek L. Secure multiparty computations on bitcoin. In *2014 IEEE Symposium on Security and Privacy*, 2014 May 18 (pp. 43–458). IEEE.

Ball M, Malkin T, Rosulek M. Garbling gadgets for Boolean and arithmetic circuits. In *Proceedings of the 2016 ACM SIGSAC Conference on Computer and Communications Security*, 2016 Oct 24 (pp. 565–577). ACM.

Bayer D, Haber S, Stornetta WS. Improving the efficiency and reliability of digital time-stamping. In *Sequences Ii*, 1993 (pp. 329–334). Springer, New York, NY.

Bellare M, Hoang VT, Keelveedhi S, Rogaway P. Efficient garbling from a fixed-key blockcipher. In *2013 IEEE Symposium on Security and Privacy*, 2013 May 19 (pp. 478–492). IEEE.

Bellare M, Hoang VT, Rogaway P. Foundations of garbled circuits. In *Proceedings of the 2012 ACM Conference on Computer and Communications Security*, 2012 Oct 16 (pp. 784–796). ACM.

Ben-David A, Nisan N, Pinkas B. Fairplay MP. A system for secure multi-party computation. In *Proceedings of the 15th ACM Conference on Computer and Communications Security*, 2008 Oct 27 (pp. 257–266). ACM.

Bentov I, Kumaresan R. How to use bitcoin to design fair protocols. In *Annual Cryptology Conference*, 2014 Aug 17 (pp. 421–439). Springer, Berlin, Heidelberg.

Bitcoin Cash. 2017. https://www.bitcoincash.org/

Buterin V. What is Ethereum? *Ethereum Official webpage*. Available: http://www. ethdocs. org/en/latest/introduction/what-is-ethereum. html. 2016.

Emercoin Blockchain. https://emercoin.com/en/emercoin-blockchain.

Emercoin. 2013. https://en.bitcoinwiki.org/wiki/Emercoin

Eris: The smart contract application platform. https://erisindustries.com/, accessed on 28-04-2019.

Geisler M. Viff: Virtual ideal functionality framework. 2007. Homepage: http://viff.dk. 2007.

Grigg I. Eos-an introduction. White paper. https://whitepaperdatabase. com/eos-whitepaper. 2017 Jul 5.

Haber S, Stornetta WS. How to time-stamp a digital document. In *Conference on the Theory and Application of Cryptography*, 1990 Aug 11 (pp. 437–455). Springer, Berlin, Heidelberg.

Halford R. Gridcoin: Crypto-currency using Berkeley open infrastructure network computing grid as a proof of work. May 2014. https://assets.ctfassets.net/sdlntm3tthp6/resource-asset-r344/bf2964ae0d86569997ce03fa95b89f55/e5a85142-f4cb-4c05-9078-271408117cf3.pdf

Hoskinson C. Why are we building Cardao. White paper. June 2017. https://cardano.org/why/

Huang Y, Evans D, Katz J, Malka L. Faster secure two-party computation using garbled circuits. In *USENIX Security Symposium*, 2011 Aug 8 (Vol. 201, No. 1, pp. 331–335).

Hyperledger Blockchain. https://www.hyperledger.org/, accessed on 28-04-2019.

Iansiti M, Lakhani KR. The truth about blockchain. *Harvard Business Review*. 2017 Jan 1; 95(1):118–27. https://hbr.org/.

Innocent AA, Prakash G, Sangeeta K. FastGarble – An optimized garbled circuit construction framework. *International Journal of Grid and Utility Computing*. In Press. https://www.inderscience.com/info/ingeneral/forthcoming.php?jcode=ijguc

Innocent AA, Prakash G. Blockchain Applications with Privacy using Efficient Multiparty Computation Protocols. In *2019 PhD Colloquium on Ethically Driven Innovation and Technology for Society (PhD EDITS)* 2019 Aug 18 (pp. 1-3). IEEE.

Innocent AA, Prakash G. Universal gates on garbled circuit construction. *Concurrency and Computation: Practice and Experience*. 2019a Sep 10; 31(17):e5236.

Innocent AA, Sangeeta K. Secure two-party computation with AES-128: Generic approach and exploiting specific properties of functions approach. In *The Fifth International Conference on the Applications of Digital Information and Web Technologies (ICADIWT 2014)*, 2014a Feb 17 (pp. 87–91). IEEE.

Innocent AA, Sangeeta K. Secure two-party computation: Generic approach and exploiting specific properties of functions approach. *Journal of Information Security Research*. 2014b; 5(1):19–27.

Katz J, Ranellucci S, Rosulek M, Wang X. Optimizing authenticated garbling for faster secure two-party computation. In *Annual International Cryptology Conference*, 2018 Aug 19 (pp. 365–391). Springer, Cham.

King S, Nadal S. PPcoin: Peer-to-peer crypto-currency with proof-of-stake. *Self-Published Paper*, 2012 Aug 19; 19.

Kolesnikov V, Mohassel P, Rosulek M. FleXOR: Flexible garbling for XOR gates that beats free-XOR. In *Annual Cryptology Conference*, 2014 Aug 17 (pp. 440–457). Springer, Berlin, Heidelberg.

Kolesnikov V, Schneider T. Improved garbled circuit: Free XOR gates and applications. In *International Colloquium on Automata, Languages, and Programming*, 2008 Jul 7 (pp. 486–498). Springer, Berlin, Heidelberg.

Kondi Y, Patra A. Privacy-free garbled circuits for formulas: Size zero and information-theoretic. In *Annual International Cryptology Conference*, 2017 Aug 20 (pp. 188–222). Springer, Cham.

Kosba A, Miller A, Shi E, Wen Z, Papamanthou C. Hawk: The blockchain model of cryptography and privacy-preserving smart contracts. In *2016 IEEE Symposium on Security and Privacy (SP)*, 2016 May 22 (pp. 839–858). IEEE.

Lee C. Litecoin-open source p2p digital currency. 2011. https://litecoin.org/

Lindell Y, Pinkas B. A proof of security of Yao's protocol for two-party computation. *Journal of Cryptology*. 2009 Apr 1; 22(2):161–188.

Mazieres D. The stellar consensus protocol: A federated model for internet-level consensus. Stellar Development Foundation. 2015 Apr; 32. https://www.stellar.org/papers/stellar-consensus-protocol.pdf

Merkle RC, Inventor; Leland Stanford Junior University, assignee. Method of Providing Digital Signatures. Patent US 4,309,569. 1982 Jan 5.

Nakamoto S. Bitcoin: A peer-to-peer electronic cash system. 2008. https://bitcoin.org/bitcoin.pdf

Naor M, Pinkas B, Sumner R. Privacy preserving auctions and mechanism design. *EC*. 1999 Nov 3; 99:129–139.

Noyes C. Blockchain multiparty computation markets at scale. 2016. https://www.semanticscholar.org/paper/Blockchain-Multiparty-Computation-Markets-at-Scale-Noyes/fb7be4b7fd591da13380b350a19b2a3ed20a0323

Patra A. Lecture Notes – CSA E0 312: Secure Computation. 2015. Retrieved March 9, 2016, from http://drona.cas.iisc.ernet.in/~arpita/SecureComputation15.html

Pinkas B, Schneider T, Smart NP, Williams SC. Secure two-party computation is practical. In *International Conference on the Theory and Application of Cryptology and Information Security*, 2009 Dec 6 (pp. 250–267). Springer, Berlin, Heidelberg.

PlatON: A high efficiency trustless computing network. https://www.platon.network/technology, accessed on 25-11-2019.

Ripple. https://ripple.com/, accessed on 28-04-2019.

Sánchez DC. Raziel: Private and verifiable smart contracts on blockchains. arXiv preprint arXiv:1807.09484. 2018 Jul 25.

Szabo N. Formalizing and securing relationships on public networks. *First Monday*. 1997 Sep 1; 2(9).

USD Coin: A cryptocurrency with a stable price. https://www.coinbase.com/usdc

Wanchain – Building super financial markets for the new digital economy. 2019. https://wanchain.org/files/Wanchain-Whitepaper-EN-version.pdf, accessed on 5-12-2019.

Wang X, Ranellucci S, Katz J. Authenticated garbling and efficient maliciously secure two-party computation. In *Proceedings of the 2017 ACM SIGSAC Conference on Computer and Communications Security*, 2017 Oct 30 (pp. 21–37). ACM.

Wang XA, Xhafa Xhafa F, Ma J, Cao Y, Tang D. Reusable garbled gates for new fully homomorphic encryption service. *International Journal of Web and Grid Services*. 2017 Feb 6; 13(1):25–48.

Wang XA. Toward construction of efficient privacy preserving reusable garbled circuits. In *International Conference on P2P, Parallel, Grid, Cloud and Internet Computing*, 2016 Nov 5 (pp. 81–92). Springer, Cham.

Willett JR, Hidskes M, Johnston D, Gross R, Schneider M. Omni protocol specification (formerly Mastercoin). White Paper, accessed Jan 2016; 28.

Yao AC. Protocols for secure computations. In *Foundations of Computer Science*, 1982. SFCS'08. 23rd Annual Symposium on (pp. 160-164). IEEE.

Zahur S, Rosulek M, Evans D. Two halves make a whole. In *Annual International Conference on the Theory and Applications of Cryptographic Techniques*, 2015 Apr 26 (pp. 220–250). Springer, Berlin, Heidelberg.

Zhou L, Wang L, Sun Y, Ai T. AntNest: Fully Non-interactive secure multi-party computation. *IEEE Access*. 2018 Nov 28; 6:75639–75649.

Ziegeldorf JH, Grossmann F, Henze M, Inden N, Wehrle K. Coinparty: Secure multi-party mixing of bitcoins. In *Proceedings of the 5th ACM Conference on Data and Application Security and Privacy*, 2015 Mar 2 (pp. 75–86). ACM.

Zyskind G, Nathan O, Pentland A. Enigma: Decentralized computation platform with guaranteed privacy. arXiv preprint arXiv:1506.03471. 2015 Jun 10.

10 Digital Linked Information System Using Blockchain Technology
Overwhelming Information Silo

B. Akoramurthy
Velammal Institute of Technology

T. Ananth Kumar
IFET College of Engineering

CONTENTS

10.1 INTRODUCTION

Data is the heart and processed data is the heartbeat in the working of a software system. From data to Big Data, the growth has witnessed a swift rise. With the evolution of Internet, the generation of data and their transformation to information have been rapid. Handling and managing surplus information have however caused a challenge of identifying and utilizing the right information at the right time in various sectors. Corporate sectors face the maximum issue in dealing with information as their communication and flow among various departments leads to the major problem of stagnation and overhead information leads to "Information Silo". Several conventional factors such as vendor diversity, mergers and acquisitions, and migration from legacy system are causes of information silo problem in corporate sectors. In addition, technological advancements in the recent years have forced information storage to move onto cloud and with the sophisticated multi-cloud communication and sharing. Despite such evolutions and diversifications in handling the problem, the complexity in complete elimination of information silo is still daunting as the root cause of it is multidimensional. Information exists with its own priorities, values, and validity, and hence sharing of it is crucial. It is generated dynamically, and the issue in handling it lies in both the communication and storage. Information sharing cannot just be an open-access model since the shared information may be of a higher value; hence, the level of information access is also to be considered. The level of control over the information sharing may be high with limited access or with a higher access but limited control, or at times access with the right control. Each department in a corporate setup has its own means of sharing/communication and hence achieving systematic information flow is an exigent task. Moreover, the task of identification and sharing the right information to the right people at the right time is a strenuous task, and it requires expertise with complete understanding of the system. There are several hitches in information sharing. For example, the research and development department in an organization selectively shares information that may be essential to marketing department. At times, nonrepudiation of information sharing too poses a challenge to the management. Though the information may be sent, the owner of the information may deny sharing the information that may lead to several management issues as well. Such an issue may not only be due to internal communicators with the corporate but also due to the external security threats. Addressing the communication issue by overcoming the information silo is one research area that has been attempted by many researchers. Several upcoming technologies have attempted to resolve the issues arising out of the information silo problem or at times purging information silos as well.

Mostly real-time systems usually have enormous interactions and multiple-type entities (MTE), such as human–social media activities, major communications and its computer systems, and typical biological networks. Such a system can be called heterogeneous information system, without its loss of generality, where its large number of interactions adds up to interrelated or connected networks. Hence, this plays a key role in present information infrastructure and everywhere. This work utilizes the concept of heterogeneous information network (HIN) because of its basic concept that mines a hidden pattern and analyze through linking several relations from the given network. HIN has also several other applications such as graph mining, network science, social network analysis, and web mining.

Some of the basic assumptions of the contemporary networks are based upon the type of objects and their links between them. If the assumption is based on the same type of objects, i.e., on a landline network, for example, each node here is telephone, does the same function as any other telephone, and users tend to buy telephones for similar reasons. These telecommunications networks are often considered as homogeneous in nature. Another good example is the author and coauthor network. In other words, absence or no consideration of more types of relations among the same type of objects can be called as homogeneous networks, where no new patterns or links or relationships can be found.

Generally, in any modern organization, data generation will not only be from a single entity or component, rather it will be from several entities. Those entities are assumed here as real-time systems, where we modeled them as HIN [27,29,32], just like in smartphones where the photos are grouped together via date, place, and time. Though the homogeneous networks are widely used because of its simplicity and less complex in nature, but by using HIN, one can have a variety of data patterns and semantics in objects. Also, HIN fuses data effectively that forms fruitful information, which is a novel development in the field of data mining. This chapter presents the basic concepts of information and information systems. Also, this chapter clearly comes out with the information identification and the formation of HIN that is needed to eradicate information silo [24,34].

10.2 BACKGROUND

Information silo problem has been analyzed for various sectors. In the work of Miller and Tucker, the silo analysis is based on the exchange of health data. This problem is eliminated by making decisions on what kind of data is to be exchanged [1]. The work of Glaser puts forward the concept of interoperability as the key in eliminating information silos in healthcare [2]. Though there is efficiency in terms of data exchange, it is also to be noted that there are security and privacy concerns when we consider interoperability as the solution to information silos. The problem of information silo in general can be eliminated by providing novel solution or mechanisms, but the challenge lies in the espousal of such strategies by the users. Hence, the solution provided must take into consideration the users who will be working on such systems. Information silos in financial sectors are most crucial as large, and information of immense importance is to be communicated. The problem with respect to the financial sectors is that neither can all information be utilized nor only specific information may be utilized as the financial data may be dependent. In their article, Phillips, Watson, and Willis analyze the problem of information silo in finance and suggest the utilization of an integrated reporting that provides a comprehensive view of information pertinent to the company, its capitalization and strategy [3]. It is to be noted that the problem of information silos has been handled in specific problems, but a generalized framework for eliminating this problem has not been given. The most common source of information silo that exists today is "Social Networking". As shown in Reference [4], social networking sites contribute to information silos in two major ways: control of information and dissemination of information by users. Decentralization that provided higher level of interaction was provided as a solution in elimination information silo problem. The drawback of the above method was the adoption of the decentralization concept by the users of social networking sites [18,24,25,28].

The advent of blockchain technology gave a newer dimension in dealing with decentralization. Blockchain technology is a recent trend that is experimented by every other industry. Blockchains have their origins in cryptocurrency platforms, in particular bitcoin, where they represent historical records of verifiable monetary stake. They were designed in the first place to solve the double-spending problem, that is, to establish consensus in a decentralized network over who owns what and what has already been spent. Blockchains are authenticated records of the history of a network's activity distributed among the users of the blockchain all around the globe. A blockchain enables secure storage of arbitrary information – in some cases, a token balance; in other systems more, complex information – within the network simply by securing a set of private keys. Blockchain technology has taken its forms with Blockchain 1.0 as cryptocurrency, Blockchain 2.0 as smart contracts, and Blockchain 3.0 as the era of business plunging into the research related to the integration of blockchain to their applications. As described in Reference [5], blockchain technology focuses on message authentication targeted toward tamper-evidence and tamper-resilience. In its most abstract form, a blockchain may be described as a tamper-evident ledger shared within a network of entities, where the ledger holds a record of transactions between the entities. To achieve tamper-evidence in the ledger, blockchain exploits cryptographic hash functions. The emergence of blockchain technology has witnessed rapid growth in terms of its applications to various sectors, ranging from cloud to supply chain management. Wright and De Filippi focus on the need to implement regulations, the governance issues, and the role of law in the deployment of blockchain technology to smart contracts [6]. In Reference [8], the application of cryptocurrency blockchain technology to validate the software license to minimize software piracy and protect software copyright was carried out. Reference [7] briefs the application of blockchain technology to financial applications, like bitcoins, insurance, etc., and nonfinancial sectors, including notary, cloud, the emerging IoT. This work enhances the scope of applying blockchain technology to emerging applications and the need for the emergence of new business models in the world of blockchain.

There has been a huge interest in mining HIN [9–11], and many proposals have shown the way to respect the differences in type that may lead to more and effective meaningful outputs. Reference [12] used a ranking-based clustering mutual enhancement method that generates various clusters or groups of multiple-type objects in it. However, the real problem is that the system or clusters will not be utilizing previous available knowledge efficiently, which shows the vague clusters [12]. In Reference [13], Yin et al. explored a social tagging graphs for HIN and classified the web objects, but in a confined way [13]. The work constructed a bipartite graph [14] between every web object and its types, which does effective classification or segmentation. Arbitrary nature of the link structure confines its application not only to generic HIN, because the method proposed is based upon the specific network schema, i.e., between tags and web objects [21,22,33].

The study of the existing works identifies the potential of the application of blockchain technology to the eradication of information silo. The security and trust achieved by blockchain technology will enable to perceive the digital information flowing through the corporate industries and streamline their flow of information and ensure the elimination of information silo problem occurring in corporate industries.

10.3   BASIC CONCEPTS AND DEFINITIONS

This section introduces some basic concepts related with HIN and compares HIN with other allied concepts with examples.

10.3.1   BASIC DEFINITIONS

An information network can be a generalization of any real world or system that comprises people, smart gadgets like mobile phones, two or more computer systems, methods, and functions, and frameworks that are organized to collect, process, and propagate data. It can be formally defined as follows.

Definition 1: Information Network

It is defined as a directed graph or digraph $G = (N, E)$ consisting of the set of N-nodes and E-edges, which are ordered pairs of elements of nodes, where nodes and edges are respectively called as objects and links.

There will be a large number of objects and their links between various other objects. Hence, a set of objects and their associated links can be represented in a simpler way as follows:

Object Set: O
Associated Links: L

Then its mapping function can be described as:

$$\Phi : N \rightarrow O \left(\text{Object-type Mapping Function} \right)$$

$$\Psi : E \rightarrow L \left(\text{Link-type Mapping Function} \right)$$

Every object (o) in the object set (O) will belong to its own object set, $O = \{o_1, o_2, o_3, o_n\}$
i.e. $\Phi(o) \in O$

Similarly, every link (l) in the link set (L) resembles the particular relation type of link set, $L = \{l_1, l_2, l_3, \ldots, l_n\}$

   i.e., $\Psi(l) \in L$

If one relation has two different associated links, then the links have same object type, both as starting and ending, which is ultimately termed as loop or self-loop.

Definition 2: Multiple Information Network

To find out various semantic meaning and to get different insights of pattern identification of the information network, it is necessary to define multiple information network as it is a new way of seeing HIN.

   Assume an information network as a graph. Let G_1 and G_2 be the multiple information networks. Both graphs G_1 and G_2 are said to be isomorphic to each other, if there exists a one on one correspondence, i.e., $F: V(G_1) \to V(G_2)$ such that for each pair x, y of vertices of G_1, $X, Y \in e(G_1)$ if and only if $F(X)$. $F(Y) \in e(G_2)$ [31].

   In other words, G_1 and G_2 are said to be isomorphic to each other if there is a mapping from one set of vertices to another set of vertices, which conserves or maintains adjacency property. Such a mapping can be called an isomorphism, and it is not possible to find out isomorphism in a single graph that is here a single information network [20]. Thus, we need two or more information networks to find out whether they are equal or not [23].

Definition 3: HIN

Consider there are "n" data objects types, which can be denoted as $x_1 = \{x_{11}, x_{12}, x_{13}, \ldots, x_1m_1\} \cdots x_n = \{x_{n1}, x_{n2}, x_{n3}, \ldots, x_{nm(n)}\}$; then HIN can be defined as a graph $G = (N, E, W)$, when $N = U_{i=1}^{n}(X_i)$, where $n \geq 2$; E is the collection of edges or otherwise associated links between nodes or connection between any data objects in the network; and W is the set of weighted values between the data objects connection [15,16,17]. If $n = 1$, then the rank of the graph reduces, which can be called as homogeneous information network [18,19].

10.3.2 CLASS-BASED HIN

A class-based HIN, $G = (N, E, W)$, can be defined as $G' = (N', E', W')$, where N' is a subset of N; here N' will consist of multiple data objects right from x_1 to x_n [30].

   All edges $\forall e = (X_{ia}, X_{bc})$ belong to E'; i.e., $\forall e = (X_{ia}, X_{bc}) \in E'$; in short, $E = E'$.
   Similarly, for links also,

$$W'_{(Xia, Xbc)} = W_{(Xia, Xbc)}, \text{ likewise } W = W'$$

The data object range for W and W' are almost equal to their part.

10.4 INFORMATION AND INFORMATION SYSTEMS (IIS)

Information, the processed data, is the backbone for any system. The flow of such information in an organization is the base for communication. Information may be defined as follows.

Definition 1

Information is a structure/packet if it's having the following properties:

- Contains identity of conceiver
- Keeps the integrity of an information
- Nonrepudiation of originator
- Is authentic (composed of all the above properties)
- Is legitimate

The traditional information systems used were paper based and difficult to manage. Tracing and tracking information was a tedious task. Such systems were not capable of updating information available at a faster pace. The drawbacks of the traditional information systems paved way to the evolution of computer-based information systems. Such systems constitute hardware, software, databases, and procedures to handle information. With the evolution and utilization of Internet, there is a significant increase in information creation and sharing. Information systems in the current scenario demands high level of sophistication and automation. Management information systems used to improve the business value and profit of an organization cannot be dependent on the information from within the organization alone. Market trends and customer demands need to be analyzed and forecasted for that particular organization. Such information may be available in the social networks that provide a valuable source of information for the business today, in understanding the customer demands and market trends. In order to make an efficient use of information in the current scenario, the information life cycle may be visualized as illustrated in Figure 10.1.

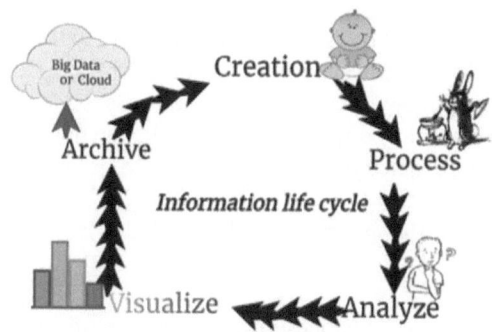

FIGURE 10.1 Information life cycle.

Information creation: Information systems in the current scenario require data from multiple sources for building an information system. The evolution of Internet has led to the need for obtaining data from various sources like social networks to understand and predict the customer demands. Such information is vital for every organization.

Information processing: In the current scenario, processing large amounts of information sources requires an information system capable of manipulating and disseminating such information at a faster pace to meet the growing demands.

Information analysis: Forecasting and prediction is the essence in analyzing information, which in turn contributes to improving the business value. In the web world of enormous information creation and processing, analysis becomes a very essential task to obtain better understanding about the available information. Information analysis requires powerful algorithms to cope with the voluminous information evolving rapidly.

Information visualization: The analyzed information is to be visualized in order to obtain better insights about the available information. Visual analytics is a powerful medium of communicating the knowledge gained though the information.

Information archival: The information needs to be archived for future utilization and here archive refers to digital archives. Identifying the appropriate information for archival is a sensitive task. At times, the need for secured information destruction may also be required. Depending on the need, information archival or destruction may be chosen.

The entire life cycle of information is encompassed in an information system. Right from information creation to its usage, archival, or destruction, it is the responsibility of the information system to monitor the flow, utilization, and distribution of information. Modern information systems demands several modifications in the existing systems due to the rapid generation of information. Information analysis and visualization has become the prime need of the business to understand the customer demands, market trends, etc. Information destruction is yet another essential task since the rapidly evolving information once used needs to be destroyed in a secure manner. The need for a highly sophisticated and automated information system is increasing day by day. Such systems should not only be able to handle information usage flow and processing but should also ensure that the issues with such systems must be purged.

Research Questions
- What kind of information is required by the organization?
- What volume of information may be required for processing?
- What is the technology used for information processing?
- What algorithms may be appropriate to handle the information available?
- Based on what parameters the information is to be chosen for analysis and visualization?
- How to identify the information that may be applicable for archival/destruction?

The answers to such research questions will aid in the efficient design of an information system. The proposed digital block information system (DIGBI) is capable of answering these questions and is able to clearly distinguish the various phases of information systems. The proposed information system considers not only the information generated within an organization but also the external information that is available to improve the business value. Despite the efficient design of an information system, there are several issues that are to be addressed by an information system. Information silo is the silent trouble that exists in several corporate sectors. Corporate industries comprising several departments intend to have high level of communication among employees. Such communication may not always be specific since organizational information will be presented in a detailed manner, which will often result in unwanted information being shared or information being stored as a bulk. At times, the essential information may not be communicated, rather unwanted information may be communicated. These are some of the common reasons for the emergence of information silo in an organization. The proposed DIGBI also uses PINSI algorithm (Purging INformation SIlo) in order to eliminate the issue of information silo. Moreover, the algorithm is also able to identify the information that may be considered for future use, thus enabling the efficient reuse of information. DIGBI is powered by the blockchain technology that is capable of handling information in a much secured manner and also enables distribution of information over nodes (computers) located at various locations. For a corporate scenario, it is complicated to identify the right information to be communicated among departments. Management issues play a major role in communication of information. Some instances include departments in corporate sectors that may not come forward to share information with other departments or due to the competitive nature among employees, essential information may not be revealed. It is not only about the communication but also the identification of the essential information that are to be archived. In an organization, the validity of information plays a major role. Though information may be archived for future use, information destruction is a very crucial task. Hence, identification of such information for destruction beyond a certain period of time is also to be considered.

An another alarming issue is the nonrepudiation of information. Information being communicated by one source is denied to be communicated. Such an issue arises when communication happens over a network. In an organizational setup where multiple users get access to the systems, ensuring proper use of information is becoming a tedious task. Despite various authentication and authorization mechanisms, information systems are not able to ensure authenticity of information. Though a computer-based information system is able to handle information, it is not capable of addressing such issues. In the current scenario, information silo along with the nonrepudiation of information has become a major threat to the corporate industries.

10.5   DIGITAL BLOCK INFORMATION SYSTEM (DIGBI)

DIGBI is capable of handling the modern information that is rapidly evolving with the help of blockchain technology. Data may be easily distributed among networked computers (nodes) that may be available in the same or different networks. As construction of blocks in a blockchain requires highly equipped computers, this information system will be capable of handling voluminous and rapidly evolving data. Data distributed over various blocks may be easily analyzed and visualized. Though the roots of this technology pertain to bitcoin, the influence of blockchain technology in the web world is still expanding. The proposed DIGBI aims to handle the issue of information silo and nonrepudiation using blockchain technology. The utilization of blockchain technology helps in identification of valid information for communication and ensures that nonrepudiation is guaranteed. The decentralized consensus mechanism in blockchain technology identifies the appropriate information for communication and ensures that no denial of information sharing is done.

Figure 10.2 clearly illustrates how nonrepudiation of information sharing can be eliminated using blockchain technology. As of the blockchain, the information is being distributed over various nodes (computers) that may be existing in the same or different network and hence enabling efficient distribution of information. Despite the distribution, the digital identification provides the most efficient means of nonrepudiation. Though authentication is still an issue to be addressed, several researches are being carried out in that direction. Figure 10.3 illustrates the flow diagram of the generic view of the DIGBI.

Figure 10.3 emphasizes the interrelationships and flow among the parameters defined. The proposed DIGBI system may be utilized either in the complete elimination of information silo or for a specific perspective, depending upon the need in an organization. Consider, for instance, that the accounts department and the finance department are working hand in hand for preparing a quotation. In this scenario, the communication that is happening among the departments will be very high as the details required have to be frequently exchanged and the information that is

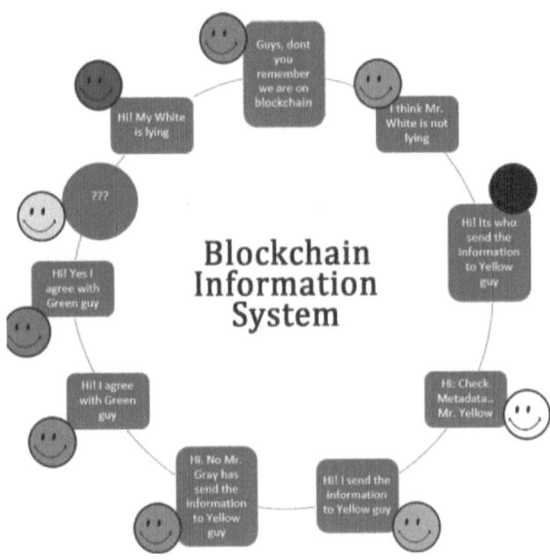

FIGURE 10.2 Elimination of nonrepudiation using blockchain.

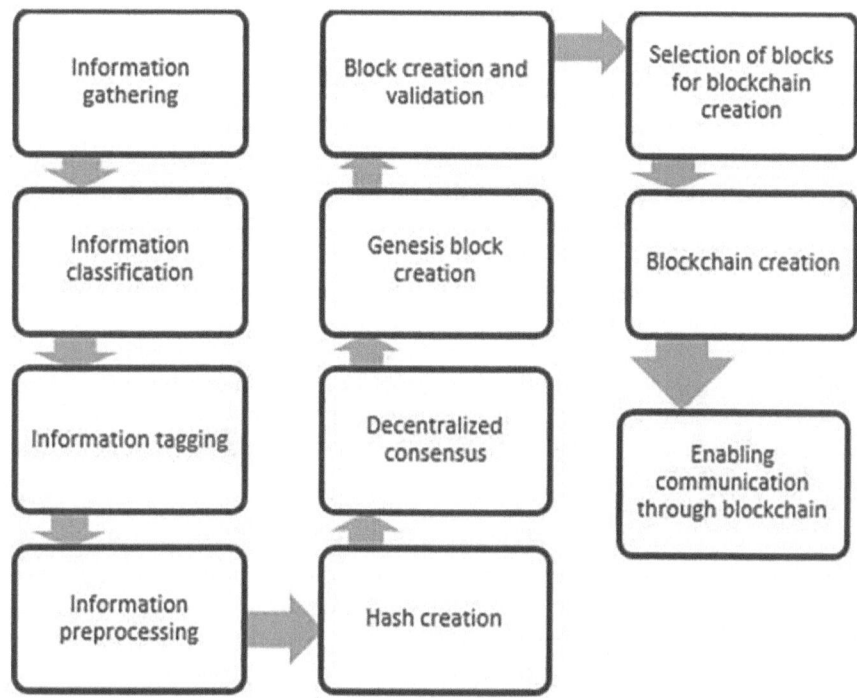

FIGURE 10.3 Workflow of DIGBI.

exchanged is highly confidential. In this case, it may be observed that in an ordinary corporate sector, the information may be flowing over a secured network and the data may be exchanged through mail servers or through the intranet facilities that may be available in the organization. Taking into account the sensitivity of the information that is being communicated, there may be leakage of information.

10.5.1 INFORMATION GATHERING

The information gathering phase focuses on the collection of information that is essential for communication in the organization. The information may be for a specific communication purpose, or it may be the entire information that the organization may be generating. The information may also be generated dynamically and that has to be added to the information repository of the DIGBI. The proposed DIGBI will be capable of handling both static and dynamic information. Such information is not restricted to the information created within the organization, it also includes the information that is gathered from the social media. It is essential for various operations on information that include analysis, visualization, etc.; in such cases, the dynamic information that is gathered is voluminous and will be generated rapidly. The voluminous and variety of information that is being gathered needs to be classified based on the domain/department.

10.5.2 Information Classification

The information gathered from various sources cannot be used directly for communication. This is because of the time taken for identifying the right information that is to be retrieved from the database. Classification of information enables easier retrieval and utilization of information. As with the existing information systems, there is less effort taken on filtering and streamlining the information that is to be communicated. The DIGBI classifies the information gathered based on the domain/departments in the organization. It does not just group the information and map it to a single domain to appropriate domains based on their utilization. For example, if the information generated belongs to the marketing department, then the same information may be useful for sales as well as accounting department. This task of information classification eliminates the problems in communicating information among various departments. In case of information to be prioritized or in case of confidential information, it may be grouped under a separate class and may be used by the concerned individuals.

In this system, let us assume that X represents the information gathered, D represents the department, i represents the number of information, and j represents the department ID. The gathered information about the organization considered is represented as X_i and the information gathered from various departments is represented D_j. Initially the gathered information is classified based on the department ID and those having higher priority that are to be communicated are being obtained.

10.5.3 Information Tagging

The classified information may be further segregated based on various parameters, depending on the communication and utilization in the organization. Parameters may be chosen dynamically as well. As the information may even be obtained from social networks to understand the customer trends, parameters may also be varied dynamically. Consider, for example, the parameters chosen are the origin, validity, and value of information. Every information will have an origin, i.e., the department from which the information has been generated, the purpose for generation, the concerned person who was responsible for the generation of the information, etc.; similarly, every information will have its own value, depending on where the information can be utilized. This parameter is crucial as this describes the sensitivity of information. The validity of information refers to how long the information will be valid. Similar to these, various other parameters may be chosen, depending on the need. Dynamic parameters may also be chosen. The chosen parameters are tagged on to the information and are kept ready for preprocessing. The task of information tagging aids in the elimination of information silo by identifying unwanted information.

10.5.4 Information Preprocessing

Information preprocessing makes the information ready for inclusion into the block. This task ensures information silo elimination by applying PINSI algorithm, which is illustrated below. This is carried out in order to ensure that the information is to be utilized at the right place and that only appropriate information is communicated among departments. In the congregating information, the information from all the departments of a particular organization are accumulated department-wise.

The backtrack function is responsible for imposing and verifying constraints (parameters chosen), which are the basic factors that are to be ensured for any organization.

Algorithm PINSI

```
Function conginfo()
For j=1 to n
      For i=1 to n
              Classify Xi based on Dj
              Prioritize Xi based on Dj
      End for
End for
Function backtrack()
For i=1 to n
      For j=1 to n
              a= Ownership
              b=Validity
              c=Utilization
              d=
              .
              .
              .
      end for
end for
function constraintsatisfaction()
      for i=1 to n
      check if c in Dj
      end for
function consistency()
      for i= 1 to n
      if above functions are called
              else if a,b,c are regenerated
              then call purge()
      end for
function purge()
      Call archive() or destroy()
```

If the information satisfies the algorithm, then the particular information is considered for the block creation. For the identified information, the corresponding hash is being created, which is the base for the creation of the block.

Hash Creation
The hash creation is used for the following purposes in a blockchain:

- To create a hash pointer that acts as a link to the previous block in a blockchain

- To avoid tampering of information that is present within a block
- To ensure security of information present within a block
- To store huge amount of information in a simple way

With the help of hashing, blockchain may be visualized as a linked list. Each of the blocks in a blockchain will consist of a block header that will hold information about the hash pointer of the previous block. This pointer creates a link with the previous block in the blockchain. Thus, each time a block is created, a hash pointer is created and a link to the previous block is also established using it.

Hashing concept is also used to identify if tampering is done to the data in the blockchain. Every information that is considered is hashed, and hence in the communication link, if there is any modification, it can be easily detected. Moreover, the capability of holding voluminous information is also simplified with the use of hash wherein for an input of varying length, an output of a fixed length can be achieved.

10.5.5 DECENTRALIZED CONSENSUS

Once the information in the information repository is hashed, the consensus process starts wherein the communication link achieves agreement on the global state of the information that is considered without any central authority and hence the term "Decentralized". As discussed earlier, in a corporate scenario, there may be several management issues that may arise and hence sharing of essential information may be prohibited. Such cases may be completely eliminated by utilizing the distributed consensus phase. After reaching the distributed consensus phase, the information may be aggregated into a block.

10.6 GENESIS BLOCK CREATION

The genesis block is the first block in the blockchain. It acts as a common ancestor for all the blocks in the blockchain. Every node in the blockchain will be aware of the block structure of the genesis block's hash structure and its details like the time of creation, etc.; the genesis block is created by gathering the essential information for communication with all its associated parameters. The information in the block is encoded and hence cannot be modified. Each time the information is accessed, every node in the blockchain is aware of it, and if any modification is done to that particular information, then that will be discarded. If the need for the modification of information arises, the process of decentralized consensus has to be approved.

10.6.1 BLOCK CREATION AND VALIDATION

Further blocks are being created with the genesis block that takes into account the need for the communication and enables to choose the right information. To create a new block, the information must be validated against the parameters and once identified to be valid, it can be considered for contribution toward the creation of a block.

10.6.2 SELECTION OF BLOCKS FOR BLOCKCHAIN CREATION

Based on the flow of information, a single blockchain or multiple blockchains of information may be created. Once the blockchain is created, further communication that happens in the organization may be easily brought into the blockchain as the context of the communication will be carried on by the parameters that are required for the creation of a block in the blockchain.

Figure 10.4 presents a simplified view of the information blockchain. It can be visualized as a linked list consisting of information. The information in the block is tagged with various parameters and since the blockchain requires highly configured nodes, large amount of information may be incorporated within the nodes. This enables the applicability of this information system to any organization. The task of handling voluminous information, the level of security provided to the information, elimination of information silo, and handling nonrepudiation of information are some of the major advantages in making use of the DIGBI.

10.7 PROOF OF CONCEPT

The proposed system that aims to eliminate information silo and ensure nonrepudiation of information can be proved, as given below. The information flow if regulated by ensuring that the right information is communicated to the right person at the right time, then it is the first step in the eradication of information silo. In the proposed system, the information is hashed and only the appropriate information is chosen with the help of parameters. As discussed in the previous section, parameters may be chosen by the user, depending upon the context of communication that will take place [26]. In the following proof, the parameters that are considered are authenticity and legitimacy as they are the most important parameters that talk about the origin of information. The proof illustrates that if the information is legitimate, then it is authentic and hence can be trusted for inclusion in the block. This is how the information is being validated before inclusion in the blocks. Several parameters may be included for this purpose, and a chi-square test may also be performed for the selection of the appropriate parameters for the creation of a block. In the proposed work, the origin of the information is given a higher priority, as for the creation of block the origin of the information has to be first verified and it is to be ensured that

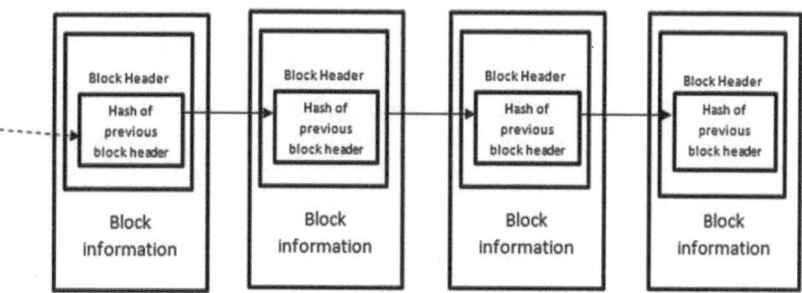

FIGURE 10.4 Simplified information blockchain.

the originator of the information can in no way deny that the information is not being sent by him/her.

Lemma 1:
   "I" is the information that is legitimate and authentic.

Proof:

Let I_{al} be the information that is both authentic and legitimate.
As a first step, decompose the information (I_{al}) into authentic information (I_a) and legitimate information (I_l):

$$I_{al} = I_a + I_l \tag{10.1}$$

To Prove:
   Case1: Information is authentic (I_a):

$$I_a = (I_r \times I_o) + I_i \tag{10.2}$$

where
I_r is the nonrepudiation of information
I_o is the identity of the originator
I_i is the integrity of information
Case 2: Information is legitimate (I_l) or I is I_i:
$I = I_i$
whereas $I = P(d)$: processed form of data:

$$\sum_{i=0}^{n} I = i_0 + i_1 + i_2 + \cdots + i_n, \quad n \geq 0 \tag{10.3}$$

Lemma 2:
   If I_0 is 1, then I is always I_a.

Proof:

Since it is known that information originates always from board of members, it will get approved by a higher authority, i.e., 1.
Though there are "n" number of information (which is like rivers originating from different places but clubbed together finally at sea as one), refined information gets aligned to the particular organization framework of information. This shows that information is ultimately legitimate:

$$I = Il = I_o \ (\text{only when o} = 1) \tag{10.4}$$

By Lemma 1, $Il = I_a$. Hence, the theorem.

10.8 RESULTS

Hypothesis

H_0: *Whether the observed value of digital linked information system meets the expected value of any business organization?*

In order to prove that hypothesis to check the proposed information system based on blockchain technology fits in any organization, we deployed test of goodness-of-fit since we had one nominal variable and small sample size; we would like to understand whether the number of observations in each classification fits a hypothetical expectation.

The table (Table 10.1) was created with the bar plot function in the R Graphics Package to produce a meek plot. The observed proportions were first calculated and then those results were copied into a matrix format that was produced by above for plotting. We labeled that matrix as ISM, often called an "R × C table", where R is the number of rows and C is the number of columns.

The graph is shown in Figure 10.5. It shows the package *ggplot2*, and uses a data frame instead of a matrix. The data frame is named Goodness fit. For the defined data frame, the code actually calculates confidence intervals, which will be added to the data frame. This code can be gamboled if the values are determined manually and can be inserted into a data frame named Goodness fit from which the plot could be generated. Sometimes it is necessary that factors need to be in level of order, which should specified for *ggplot2* to put them in the right order on the plot; else, R will alphabetize levels. As the expected value for null hypothesis is within a range 0.5–0.6, our hypothesis is accepted as defined conditions met.

Figure 10.5 shows a bar plot of Fitness for Organization vs. Information Systems. Error bars indicate 95% confidence intervals for each observed proportion.

TABLE 10.1
Information System Matrix(ISM)

Information Systems	Value	Count	Total	Proportion	Expected
Digital IS[a]	Observed	70	156	0.4487	0.54
Digital IS[a]	Expected	54	100	0.54	0.54
SDLC-based IS[a]	Observed	79	156	0.5064	0.40
Software development life cycle–based IS[a]	Expected	40	100	0.40	0.40
Traditional IS[a]	Observed	3	156	0.0192	0.05
Traditional IS[a]	Expected	5	100	0.05	0.05
MIS[b]	Observed	4	156	0.0256	0.01
MIS[b]	Expected	1	100	0.01	0.01

[a] Information Systems
[b] Management Information Systems

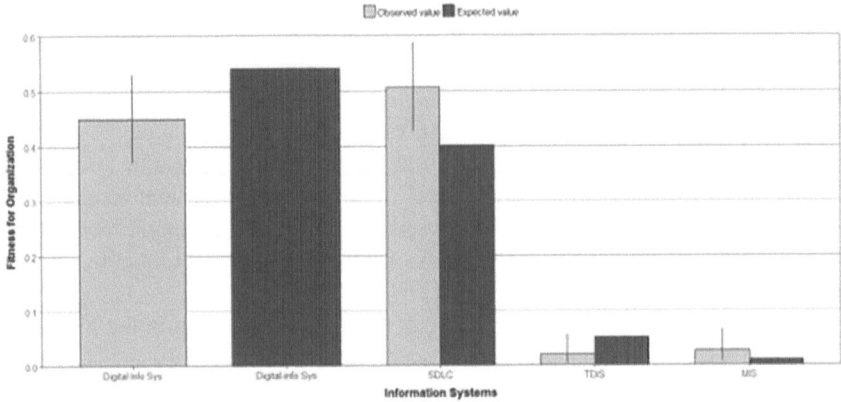

FIGURE 10.5   Fitness of DIGBI to an organization.

10.8.1   CHI-SQUARED TEST

Data: observed data

X-squared = 3.4532, degrees of freedom = 1, p-value = 0.5568

Since the p-value is higher than alpha level (0.5568 > 0.467), the null hypothesis is accepted, which ensures the digital block-based information system holds a greater hand when compared to other existing Information Systems.

Also, our X-squared value signposts how our observed data can be applicable to or fit into any organization.

10.8.2   CHI-SQUARED POWER CALCULATION

w = 0.1

The variable w is *effect size*, which is used to define how much the alternate hypothesis affects the given data samples comparably with null hypothesis; generally, the value should be low as much as possible, in order to accept the null hypothesis. Here the effect size is 0.1, which means we had 0.1% of effect in observed data in the alternate hypothesis, which stands as proof for definition of null hypothesis shall be accepted.

n = 963.4689 (n is the number of observations)

Degrees of freedom = 2 significant level = 0.05

Power = 0.80

In order to understand how much the proposed system has consistency in achieving the power over other information system in any organization, we deployed the power test based on chi-square Goodness for Fit. The power value we attained is 80%. This value shows that the efficiency in incorporating digital linked information system comparably with other information systems is high enough, which boosts up the information flow in legitimate manner, which defines the path flow incredibly as the path demarcation has been grounded upon blockchain technology.

10.9 CONCLUSION

Information silo has witnessed a gradual rise in several organizations with the evolution of digital information. The management and handling of such information is a complex task for the employees in the organization. Communication, being the base in any organization for its smooth functioning, demands information to be on the move all the time. Such information movement is affected by several factors such as unwanted information being shared, essential information not being shared among the departments due to various management reasons, highly competitive environments leading to malpractices in information handling, etc., The most alarming and rising problem in corporate sector is the challenge in ensuring nonrepudiation of information. The proposed work aims to handle information silo and ensure nonrepudiation of information using the popular blockchain technology. Information that flows in the organization is identified and corresponding blocks are created based on the communication link. The decentralized consensus process ensures nonrepudiation of information and other techniques involved in blockchain including hashing ensure tamper-proof information. The creation of blocks ensures that information dumping cannot happen and the information is utilized efficiently. The proof of concept makes it evident that the appropriate parameter selection in the identification of information for its inclusion in the block of a blockchain contributes to the elimination of information silo to a greater extent. The vast application of the blockchain technology to various domains is the major strength of the proposed system, making it efficient in handling the information silo problem. The results using the goodness of fit test clearly indicate that the application of block information system to an organization will be efficient compared to the various other information handling systems. The complete automation of the proposed block information system may be achieved with the help of artificial intelligence, which may be a future enhancement to the proposed information system.

REFERENCES

1. A. R. Miller and C. Tucker, "Health information exchange, system size and information silos," *Journal of Health Economics*, vol. 33, 2014, pp. 28–42.
2. J. Glaser, "Interoperability: The key to breaking down information silos in health care," *Healthcare Financial Management*, vol. 65, no. 11, 2011, pp. 44–49.
3. D. Phillips, L. Watson, and M. Willis, "Benefits of comprehensive integrated reporting: By standardizing disparate information sources, financial executive can eliminate the narrow perspectives of the elephant and the blind man parable—and "see" beyond merely information silos or reports," *Financial Executive*, vol. 27, no. 2, 2011, pp. 26–31.
4. C. M. A. Yeung, I. Liccardi, K. Lu, O. Seneviratne, and T. Berners-Lee, "Decentralization: The future of online social networking," in *W3C Workshop on the Future of Social Networking Position Papers*, vol. 2, 2009, January, pp. 2–7.

5. M. Pilkington, "Blockchain technology: Principles and applications," In F. Xavier Olleros and Majlinda Zhegu, Associate Professors of Innovation Management, École des Sciences de la Gestion, Université du Québec à Montréal, Canada Research Handbook on Digital Transformations, Edward Elgar Publishing. 2016, p. 225.

6. A. Wright and P. De Filippi, "Decentralized blockchain technology and the rise of lex cryptographia," 2015. Retrieved August 1 (2015): 2017. SSRN Electronic Journal, pp. 1–58.

7. M. Crosby, P. Pattanayak, S. Verma, and V. Kalyanaraman, "Blockchain technology: Beyond bitcoin," *Applied Innovation*, vol. 2, 2016, pp. 6–10.

8. J. Herbert and A. Litchfield, "A novel method for decentralised peer-to-peer software license validation using cryptocurrency blockchain technology," in *Proceedings of the 38th Australasian Computer Science Conference (ACSC 2015)*, vol. 27, 2015, January, p. 30.

9. A. Banerjee, S. Basu, and S. Merugu, "Multi-way clustering on relation graphs," in *Proceedings of the 2007 SIAM international conference on data mining, Society for Industrial and Applied Mathematics*, 2007. pp. 145–156.,

10. R. Bekkerman, R. El-Yaniv, and A. McCallum, "Multi-way distributional clustering via pairwise interactions," in *ICML '05*, 2005, pp. 41–48.

11. B. Long, Z. M. Zhang, and X. W. P. S. Yu, "Spectral clustering for multi-type relational data," in *ICML '06*, 2006, pp. 585–592.

12. Y. Sun, Y. Yu, and J. Han, "Ranking-based clustering of heterogeneous information networks with star network schema," in *KDD '09*, 2009, pp. 797–806.

13. Z. Yin, R. Li, Q. Mei, and J. Han, "Exploring social tagging graph for web object classification," in *KDD '09*, New York, NY, 2009, pp. 957–966.

14. U. Kang, C. E. Tsourakakis, and C. Faloutsos, "Pegasus: A peta-scale graph mining system implementation and observations," in *ICDM*, 2009, pp. 229–238.

15. W. Shen, J. Han, and J. Wang, "A probabilistic model for linking named entities in web text with heterogeneous information networks," in *SIGMOD*, 2014, pp. 1199–1210.

16. M. Danilevsky, C. Wang, F. Tao, S. Nguyen, G. Chen, N. Desai, L. Wang, and J. Han, "Amethyst: A system for mining and exploring topical hierarchies of heterogeneous data," in *KDD*, 2013, pp. 1458–1461.

17. X. Yu, Y. Sun, P. Zhao, and J. Han, "Query-driven discovery of semantically similar substructures in heterogeneous networks," in *KDD*, 2012, pp. 1500–1503.

18. Q. Zhan, J. Zhang, S. Wang, S. Y. Philip, and J. Xie, "Influence maximization across partially aligned heterogeneous social networks," in *PAKDD*, 2015, pp. 58–69.

19. J. Zhang and P. Yu, "MCD: Mutual clustering across multiple heterogeneous networks," in Proc. IEEE Int. Congr. BigData, 2015. pp 762–771.

20. S. Jin, J. Zhang, P. S. Yu, S. Yang, and A. Li, "Synergistic partitioning in multiple large scale social networks," in *IEEE BigData*, 2014, pp. 281–290.

21. G. W. Klau, "A new graph-based method for pairwise global network alignment," *BMC Bioinformatics*, vol. 10, no. Suppl 1, 2009, p. S59.

22. D. Koutra, H. Tong, and D. Lubensky, "Big-align: Fast bipartite graph alignment," in *ICDM*, 2013, pp. 389–398.

23. J. Zhang and P. Yu, "Integrated anchor and social link predictions across social networks," *in*. Proc. 24th Int. Conf. Artificial Intell., 2015, pp. 2125–2131.

24. J. Wu, L. Chen, Q. Yu, P. Han, and Z. Wu, "Trust-aware media recommendation in heterogeneous social networks," *World Wide Web*, vol. 18, no. 1, 2015, pp. 139–157.

25. J. Zhang, X. Kong, and P. S. Yu, "Predicting social links for new users across aligned heterogeneous social networks," in *ICDM*, 2013, pp. 1289–1294.

26. J. Zhang, X. Kong, and P. S. Yu, "Transferring heterogeneous links across location-based social networks," in *WSDM*, 2014, pp. 303–312.

27. F. Liu and S.-T. Xia, "Link prediction in aligned heterogeneous networks," in *PAKDD*, 2015, pp. 33–44.
28. Y. Zhang, J. Tang, Z. Yang, J. Pei, and P. S. Yu, "Cosnet: Connecting heterogeneous social networks with local and global consistency," in *KDD*, 2015, pp. 1485–1494.
29. Q. Zhao, S. S. Bhowmick, X. Zheng, and K. Yi, "Characterizing and predicting community members from evolutionary and heterogeneous networks," in *CIKM*, 2008, pp. 309–318.
30. C. C. Aggarwal, Y. Xie, and S. Y. Philip, "On dynamic link inference in heterogeneous networks," in *SDM*, 2012, pp. 415–426.
31. J. M. Kleinberg, "Authoritative sources in a hyperlinked environment," in *SODA*, 1999, pp. 668–677.
32. D. Zhou, S. A. Orshanskiy, H. Zha, and C. L. Giles, "Co-ranking authors and documents in a heterogeneous network," in *ICDM*, 2007, pp. 739–744.
33. H. Deng, M. R. Lyu, and I. King, "A generalized co-hits algorithm and its application to bipartite graphs," in *KDD*, 2009, pp. 239–248.
34. C. C. Aggarwal, Y. Xie, and P. S. Yu, "A framework for dynamic link prediction in heterogeneous networks," *Statistical Analysis and Data Mining: The ASA Data Science Journal*, vol. 7, no. 1, 2014, pp. 14–33.

11 Blockchain
Next-Generation Technology for Industry 4.0

R. Rajmohan, T. Ananth Kumar,
M. Pavithra, and S. G. Sandhya
IFET College of Engineering

CONTENTS

11.1 INTRODUCTION

Cryptographic forms of money and blockchain advancements have become amazingly well known and compelling in merely less than 10 years since the start of bitcoin. In each country, the administration can either stifle or make a specific wonder or development. Concerning cryptographic forms of money, governments in various countries are reacting suddenly. The conflict among governments and group financing, interest, and the utilization of digital currencies as units of trade is one of the numerous obstacles that should be taken into account while making use of blockchain. Also, the help offered by different governments is gradually supporting the presentation and development of digital forms of money and blockchain [1], leading to decrease of expenses and increase of global exchanges. These framework redesigns have minimal influence on society and the general population, although speeding up the process of sending cash from one spot to the next. In this way, IT departments in banks are useful as it enables people to freely oversee and control their funds and communicate with these establishments. It would have been problematic if the individuals had the option of taking out the banks and their framework and going about as their banks utilized bitcoin or some other decentralized cash.

Notwithstanding the tremendous institutional selection of blockchain, with innovations blockchain has assumed a primary role in the creation of decentralized applications, which as of now disturbs an immense number of built-up enterprises. The newest utilization of blockchain and digital currency is to make efficient power vitality an effectively tradable element. Blockchain innovation or circulated records are used to encourage the prompt exchanges between two gatherings, for instance, two neighbors.

Blockchain-based design for securing current electronic health records (EHR) frameworks [2] is a promising advancement in security field. EHRs usually encompass exceedingly delicate and perilous records correlated to patients. It tracks all the events that occurred in the records to achieve data integrity for secure transactions. To solve these problems, smart agreements, classified contracts, and user record-associated contracts are realized. The classified contract involves organizing the records in a distributed ledger format with secured interventions of doctors and healthcare providers. The user record-associated contracts validate the transactions requested by the concerned miners. Blockchain-based architecture for securing EHR is implemented in the Hyperledger framework, which includes Hyperledger explorer, composer, and caliper. Compared to conventional EHR frameworks, the recommended charter with the employment of the blockchain has improved efficiency and security for storing EHRs.

Since the evaluation of bitcoin, blockchain has been applied in a real-time environment. Blockchain is completely decentralized, immutable, tamper-proof, and secure. Only the authorized user can be able to see the data in the blockchain network that provides transparency. By this, blockchain can be implemented in the KYC chain to provide decentralized data storage and transparency. Secured KYC information is shared with various banks through the blockchain network with customer's approval [3]. The government also verifies the information to check that the data provided by the customer are valid and maintenance cost is reduced by the creation of a distributed network on blockchain.

Cryptographic forms of money have been the greatest monetary crisis in the digital world. Through blockchain innovation, bitcoin offers incredible potential for sanitation and check in the agri-food division [4]. However, it is a long way from being the panacea for a scope of issues influencing business. Basically, blockchain innovation is a method for removing and sharing data over a system of clients in an open virtual space. Blockchain innovation enables clients to take a gander at all exchanges at the same time and progressively. In nourishment, for instance, a retailer would know with whom his provider has had dealings. Furthermore, since exchanges are not made in any single area, it is practically difficult to hack the data. For buyers, blockchain innovation can have any kind of effect. By perusing a basic QR code with a cell phone, information, for example, related to an animal's date of birth, consumption of anti-toxins, immunizations, and area where the domesticated animals were collected can without much of a stretch be passed on to the purchaser.

Blockchain makes food storage network progressively straightforward at an all-new level. It additionally engages the whole chain to be progressively receptive to any sanitation debacles. Huge associations, for example, Nestlé and Unilever, are thinking about blockchain advancements for that very explanation. Walmart, which sells 20% of all nourishment in the United States, has quite recently finished two blockchain pilot ventures. Prior to the utilization of blockchain, Walmart directed a follow back test on mangoes in one of its stores. It took 6 days, 18 hours, and 26 minutes to follow mangoes back to its unique ranch.

By utilizing blockchain, Walmart can give all the data on the purchaser's needs in 2.2 seconds. During an episode of sickness or sullying, 6 days is an unfathomable length of time. An organization can spare lives by utilizing blockchain innovations. Blockchain likewise permits explicit items to be followed at some random time, which would assist with decreasing nourishment squander. For example, debased items can be followed effectively and rapidly, while safe nourishments would stay on the racks and not be sent to landfills.

11.2 BLOCKCHAIN TECHNOLOGY OVERVIEW

Blockchain is a computerized, decentralized (circulated) record that tracks all exchanges occurring over a distributed system. It is an interlinked and persistently extending rundown of records put away safely over various interconnected frameworks. Thus, blockchain innovation is versatile because the system has no single purpose of powerlessness. Furthermore, each "block" is exceptionally associated with the past squares through an advanced mark, which implies that creating a change to a record without upsetting the files in the chain render the data sealed. The critical advancement in blockchain innovation is that it permits its members to move resources over the Internet without the requirement for a concentrated outsider.

The blockchain framework is created as the underlying transaction network behind the digital currency called bitcoin. In 2008, subprime emergency decreased trust in the current money-related framework. Satoshi Nakainoto composed a white paper containing the "bitcoin convention", which utilized a conveyed record and agreement to register calculations. The bitcoin convention was composed to disintermediate customizable monetary go-betweens as methods for encouraging direct P2P exchanges. Since the introduction of the Internet, there have been endeavors to make virtual monetary standards; however, those endeavors flopped due to the "twofold spend" issue, precisely the hazard that a digital resource, for example, cash can be burned through twice. The instant answer for disposing of the twofold spend issue is through the presentation of mediators of trust, for example, banks. Be that as it may, the utilization of blockchain innovation makes it conceivable to unravel the central problem of twofold spending without the requirement of middleman of trust, encouraging the exchange of advantages, for example, virtual monetary forms over the Internet, safely. This idea can be stretched out to non-money-related territories, and that is the guarantee of blockchain innovation.

An essential relationship is a contrast between the Internet and the intranet. While the innate innovation for arranged PCs continues as before, there is a significant contrast between the elements and utility related with a shut system (for example, a home system) and an open system (for example, the Internet).

In all actuality, this distinction plays out on how "hubs" are boosted to stay a piece of the system. The critical thought here is that in an open blockchain, the agreement system is based around compensating every individual member to keep up with the system. In a private blockchain, the requirement of making this motivator doesn't exist. The decentralized idea of a competent, straightforward, open record probably won't be of utility to an association or an endeavor arrange as the gatherings are known, and a degree of understanding exists about which individuals can take an interest in the system and on what sort of exchanges. The general agreement is that open blockchain functions admirably for specific applications. For example, cryptocurrency-based exchanges (bitcoin), the more prominent utilization of blockchain innovation as an endeavor arrangement, would just be conceivable with the expanded administrative control related to a private blockchain environment (Figure 11.1).

Subsequently, its various applications are developing continuously and iteratively. The blockchain procedure works as follows [5]:

- Solicitation or the exchange initially speaks online to the square itself.
- Once the data are received by the block, it again sends to the gatherings of the clients.
- Once this solicitation is received by the collections, it is examined and endorsed by all the gatherings.
- After endorsement of solicitation, the new square can now be authoritatively added to the chain. After this block is added, the cash is moved to the subsequent person.

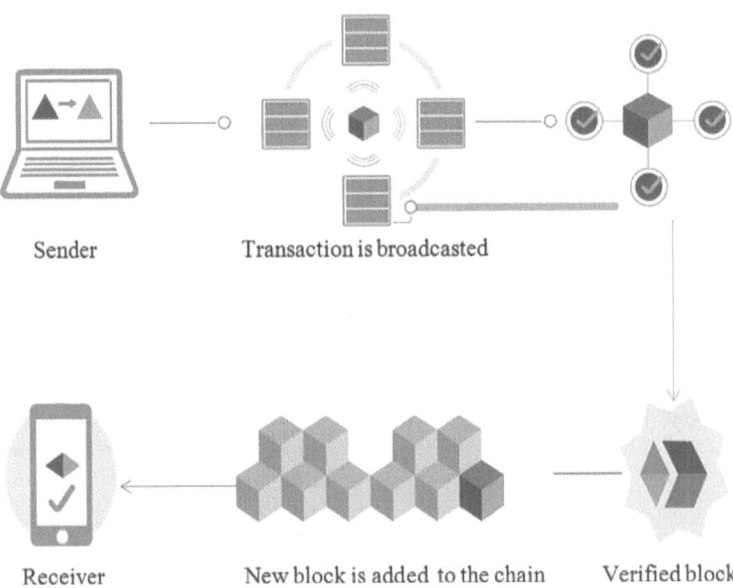

| Sender | Transaction is broadcasted |

| Receiver | New block is added to the chain | Verified block |

FIGURE 11.1 Blockchain working.

By using the advanced ID of the wallet, the amount value is moved from the sender to the collector, depending on the blockchain innovation. Miners in the system ought to confirm each exchange employing hash capacities. Wallet is found in the Mist program. The off chance to wear a current application programming interface (API) of the wallet should be maintained carefully; concerned with MyEtherWallet (MEW), a new wallet will be allocated deliberately. MEW is utilizing the cornerstone record or the secret key generated by accomplishing the above process.

A block record contains a few or entire albums of latest cryptographic money exchanges that have been entered into the new record accessed by the blocks. Every block has a resemblance to the other block that has sustained with records. Blockchain is a decentralized structure consisting of numerous levels of blocks, where every individual block contains a hash code of the previous block (Figure 11.2).

A blockchain stores data about the transactions between clients or different changes to its database: They can always be followed by a client for straightforwardness. Some members, the miners, assemble data about exchanges and arrange them in supposed blocks. Utilizing immense measures of vitality, it is ensured that the request for all trades at any point is secure and permanently archived in the de-central database of which any client has a copy on his gadget (Figure 11.3).

Mining is the way toward adding exchange records to the extensive history of the past exchange. Each exchange is approved before being added to a block. The block is made, and a block message is sent. At the point when an excavator gets another block, it is approved before taking a shot at it. When the block is finished, it is recommended to communicate it to every hub tolerating it. To add another exchange to the current blockchain, an excavator initially confirms the legitimacy of the trade. It should then settle an expensive crypto puzzle to exemplify this exchange in another block, before dispersing this square to different miners. Blockchain has some key features

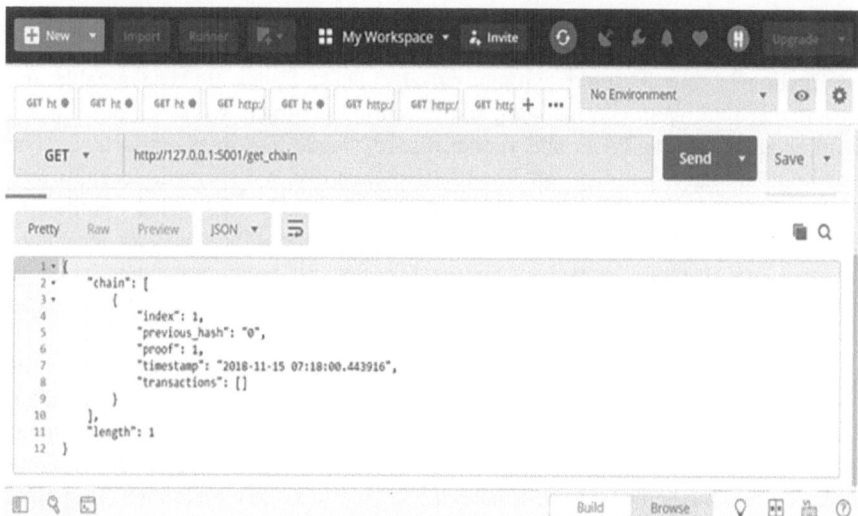

FIGURE 11.2 Block creation.

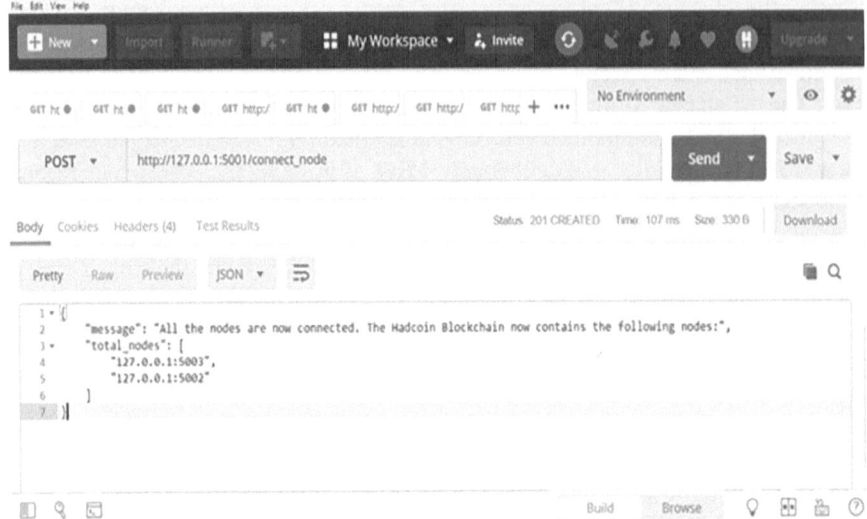

FIGURE 11.3   Mining and transaction.

regarding security, integrity, and anonymity to avoid unauthorized user validation. These features have made the blockchain a feasible solution for secure transaction; for security and privacy purposes, blockchain uses the SHA-256 (secure hashing algorithm) cryptographic algorithm: a single-way function that converts records into a 256-bit hash value for the decentralized architecture.

11.3   SMART CONTRACTS AND DIGITAL WALLETS

A smart contract [6] is a PC show intended to deliberately energize, check, or approve the course of action or execution of an understanding. Smart agreements grant the introduction of reliable trades without pariahs. These trades are recognizable and irreversible. Intelligent contracts are self-executing contracts with the states of the comprehension among buyer and vendor directly made into lines out of code. The code and the understandings contained in that exist over a dispersed, decentralized blockchain mastermind. Smart agreement award accepted trades and opinions to be done among various, puzzling social occasions without the necessity for a central force, real structure, or external approval framework. Smart agreement produces byte code using quality language, and the system is used to unscramble the advanced wallet (Figure 11.4).

A digital wallet [7] is a system that securely stores information and passwords of customer for different strategies and destinations. By using a propelled wallet, customers can complete purchases adequately and quickly with close field correspondences development. They can moreover make more grounded passwords without obsessing about whether they will have the choice to review them later. Propelled wallets can be used to identify with versatile portion systems, which grant customers to pay for purchases with their PDAs. An electronic wallet can, in like manner, be

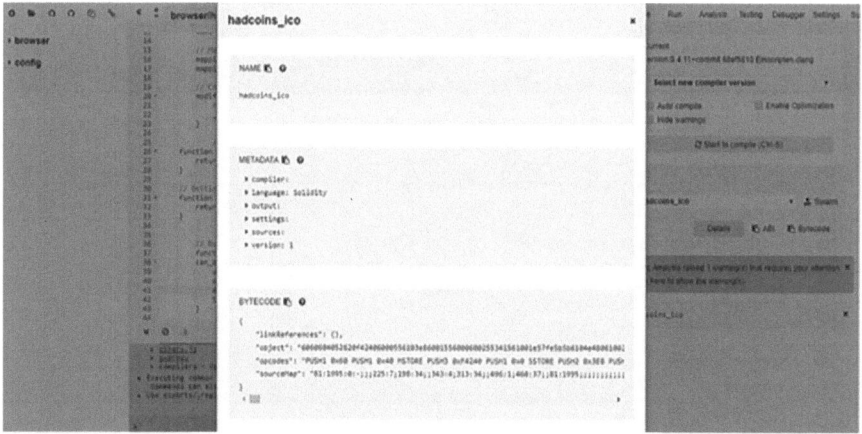

FIGURE 11.4   Smart contract creation.

used to store dependability card information and propelled coupons. A modernized portfolio is, in any case, called an e-wallet (Figure 11.5).

Ether wallets can be connected to the blockchain using the byte code. Ethereum is the development platform for blockchain operations, and Ganache is the private blockchain used by the Ethereum environment around every record concerned with the block (Figure 11.6).

Postman is an API improvement device that assists with building, test, and changes in APIs. Practically, any usefulness that could be required by any designer is exemplified right now. It is utilized by more than 5 million designers consistently to make their API advancement simple and basic. It can make different kinds of HTTP demands (GET, POST, PUT, PATCH), sparing conditions for some time in the

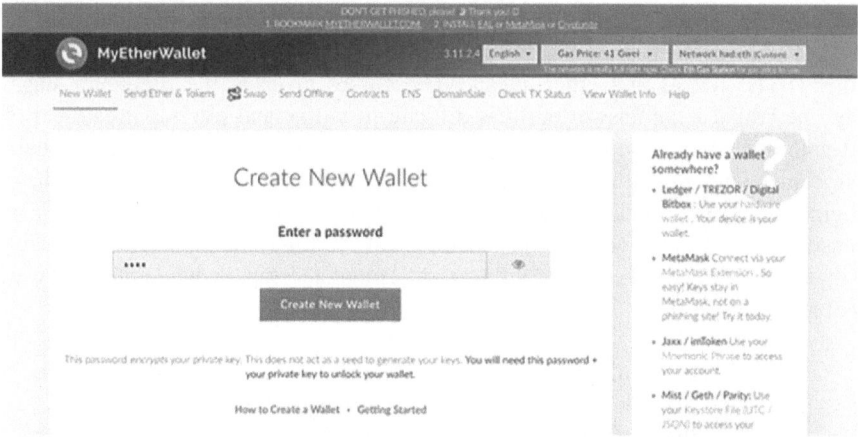

FIGURE 11.5   Digital wallet creation.

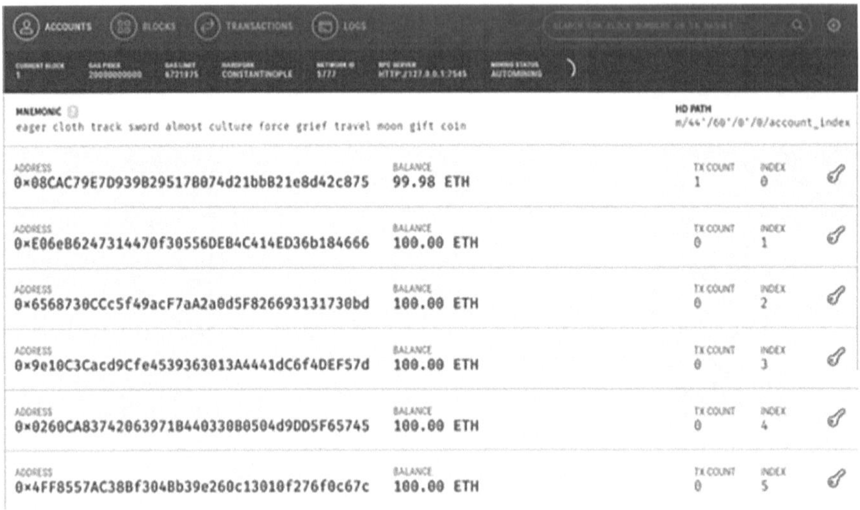

FIGURE 11.6   Fees transaction.

future, changing over the API to code for different dialects (like JavaScript, Python). HTTP Operations involved in API are as follows:

- GET: A particular asset is read by an identifier or an assortment of assets.
- HEAD: It has the principles concerned with returning headers like GET.
- PUT: A particular asset can be updated by an identifier or an assortment of assets. It is used to make an asset if the miner knows the asset identifier.
- DELETE: An identifier can remove a particular asset.
- POST: New resource creation. It is collecting all verbs for operations that should not involve with another group.

MyEtherWallet (MEW) is an unrestricted, exposed source instrument for making wallets that involved with the Ethereum platform. With regard to putting away ether (ETH) tokens and other advanced resources given at the Ethereum stage, MyEtherWallet is an answer but it is difficult to utilize and offers a great deal of adaptability. MEW is a customer using API that cooperates with the blockchain associated with the Ethereum. Although new wallets in the Internet searching engine with MEW is made without much of a stretch, it's not an online wallet. MEW consolidates the simplicity of arrangement given by online wallets while dispensing with a significant number of the vulnerabilities that accompany while putting away the assets online:

- Critical sites should be checked: It's imperative to cross-check to ensure safe site utilization, "https://" convention instead of "Http".
- Create another wallet: After it is confirmed that the visited site is right, with a welcome message titled "Make a New Wallet."

- Select a private key and enter a specific password of the wallet into the field. Keep this secret key, so save the private key in someplace for further performance.
- Download the critical store: MyEtherWallet will utilize the secret key to create a different key to store records for the wallet. Instructions need to be followed, and the alerts on sharing the unique key with others before downloading the document should be considered.
- Wallet private key will be generated after the instructions are completed. Add and print the complete information as a "paper wallet".
- Check the confirmation after receiving the private key: Approach will be made to pick who will have the option to get to the record.
- Sign in: Enter the suitable data dependent on the favored strategy for getting to the record. At this point, the wallet is supposed to open, the option to see the record address will be visible, and the parity and one of a kind QR code related to the record.
- Move the existing Ether: If it contains ether in some other record or on a digital currency trade, and then transfer the assets into another Wallet to store the ether via MEW address.
- Develop new Ether: When no ether is owned, buy ether to support the record. It cannot be obtained legitimately through MyEtherWallet.

Solidity is an agreement confined to specific language for executing smart contracts with the concerned language such as C++, JAVA, and Python. Strength is statically composed, underpins legacy, libraries, and sophisticated client-characterized types among different highlights.

11.4 KYC USING BLOCKCHAIN IN INDUSTRY 4.0

Managing the same customer information in various banks and other financial sectors creates data redundancy and high cost for maintaining sensitive information. In the traditional system, there is also a lack of security. Blockchain-based KYC verification can solve all the problems that exist in the traditional method. Using the generation of signature keys, the signature token is required to access the information of the individuals, as certain actions can perform only through the user permission. The main advantage of using blockchain is to reduce the data duplication and give an assurance to the customer for their information.

Bhaskaran proposed a data sharing method focused on two fundamental pillars: distributed consent management and double-blind (anonymous) data sharing [8]. Here, blockchain transactions rely on zero-knowledge proofs. An anonymous relationship between a service provider and a requester is proposed. It helps in building a confidential relationship between a service provider and a customer. And it is more useful for multi-party business transactions.

Shbair et al. proposed an article on blockchain orchestration and experimentation framework that it is easy to use orchestration over the grid 5000 [9]. It is a highly reconfigurable and controllable large-scale testbed to perform the blockchain framework to make the smart contract. The proposed framework can be used to study the behavior of blockchain, with specific configurations, under massive KYC client registration requests, which can be able to conduct blockchain evaluation on a large-scale environment.

Mondal et al. proposed an elective strategy for approving/getting paperwork done for budgetary administrations through the web [10]. It is a technique to limit money related to misrepresentation fabrication on an online monetary system. The proposed strategy introduced there incorporates delicate individual data called KYC data to check the genuine proprietor of the record for online money-related action. It gives better execution to confirm online payment-related action.

Wang et al. predicted Decentralized Applications based on how the blockchain works as well as how it connects with the peer-to-peer (p2p) network and provide security to the information [11]. The participants in the decentralized ledger system need to achieve consensus, an agreement upon every message being broadcast to each other.

Blockchain interfaces with the companion-to-peer arrangement and give security to the data. The members in the decentralized record framework need to accomplish accord, an understanding upon each message being communicated to one another. Open blockchains are likewise alluded to authorization of fewer blockchains as framework members need not bother about any consent before joining the system. A private blockchain approaches to control and works under a particular association. Members should be welcomed, and existing members may settle on future participants.

Previously, if a customer wanted to create an account in the bank, they had to fill a lot of personal identity documents for verifying the customer. They used to fill a lot of handwritten papers generally in the bank. They had to fill all the details for creating an account in a particular bank. The same customer had to do the same process for creating an account in another bank, which was a time-consuming process and the same details were stored in the different bank databases. If the customer wanted to create an account in 10 banks, the same process was carried out in all banks with some other additional information. It was required to maintain the customer information in all the banks; the banks did not get any revenue for this; instead, a huge amount was wasted at the cost of the end-user. All banks, more or less, got the same information from the customer that led to data redundancy. However, at present, if this information is stored and managed by the centralized organization, this also leads to data insecurity. Through centralized organization control, hackers can easily crash the server and collect or modify the information; so the existing system is less secure, do not give confidentiality to the customer's information (Figure 11.7).

The customer submits the KYC document to the respective bank. They fill all the information that is required in the report to be filled and submitted to the bank. Then the bank forward the KYC documents to the centralized intermediary after verifying the person and then it moves to the centralized database for which bank spends a lot of money as a maintenance cost of the customer information and security purposes.

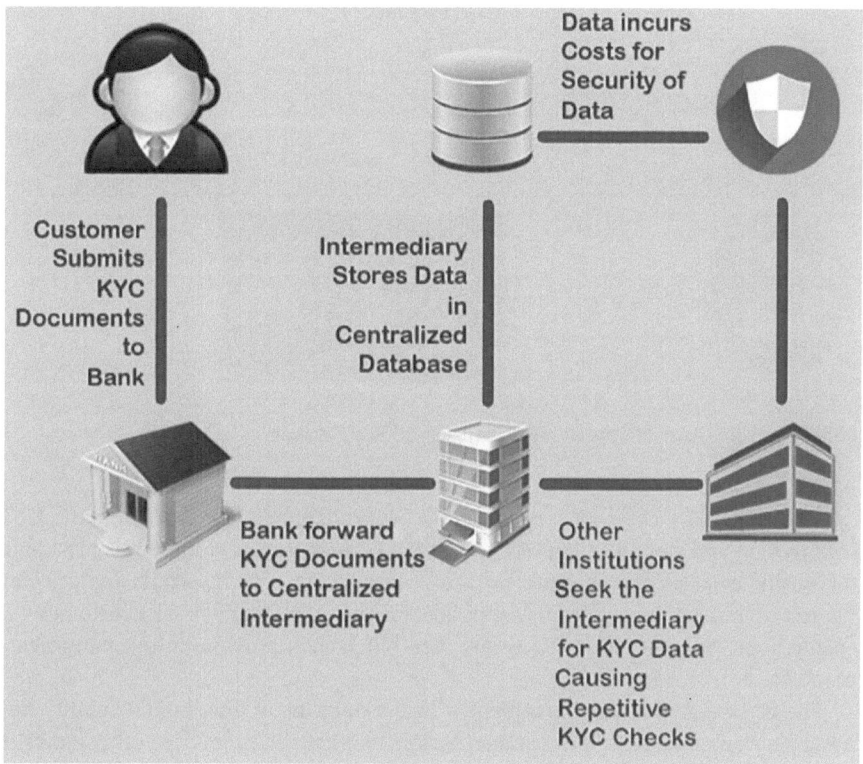

FIGURE 11.7   Existing KYC details maintenance system.

The same data being stored in the centralized database cause redundancy and wastage of money. The KYC information is unnecessarily replicated by multiple banks for creating an account. From the bank side, the banks maintain a vast amount of data in a centralized database and spent over a million for their maintenance. Such data do not give any incentive to the corresponding banks. Since the data are stored in a centralized database, they lack security. There is no insurance given to the customer's information from the bank side. The blockchain-based KYC verification system can create a block for each bank and the customer add the KYC information and store it into the blockchain network by creating account on blockchain through the customer account. They can request the bank that they need to create an account for which the details stored in the blockchain will be modified or changed only by the customer and through customer's permission. The requested bank can view the customer KYC document by giving a view request to the customer profile, and the customer may give an option of either allow or deny if the customers allow the view request, and then the blockchain provides a transparent view to the bank and government can be able to verify the KYC document that is issued by the government. The information stored in the blockchain is additional security (Figure 11.8).

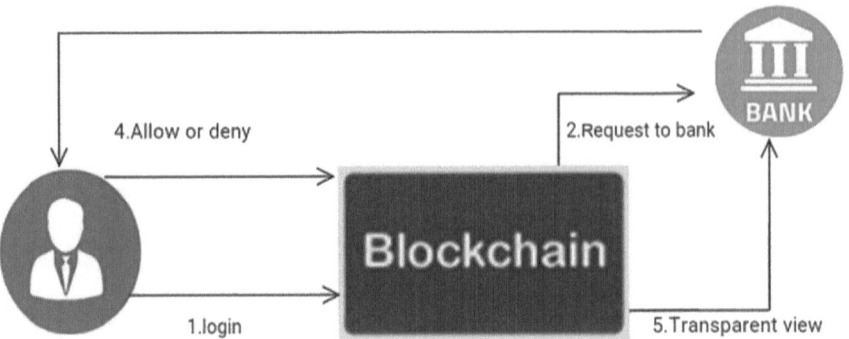

FIGURE 11.8 Architecture of blockchain-based KYC system.

The customer creates an account on the blockchain network and enters the KYC information and store it in the blockchain network as well. Through the blockchain, the customer requests the bank to create an account, and the bank gives a view request to the customer. They have options of either allow or deny. The customer will enable us to view the information and then blockchain provides a transparent view to the bank.

The bank can make an agreement with the other bank for a confidentiality purpose. Through the bank, the customer is able to create an account in other banks. If some additional information is needed in the bank, it can ask the customer directly. And only the copy of the data can be viewed by the other bank through the central bank along with blockchain; to make any changes, the updates will be notified to all other connected networks and here the network is the bank.

The customer must have an active account on the blockchain. They upload the entire documents that are needed to create an account to the corresponding bank. The government is able to verify the details that are provided by the customer and the same as issued from the government are cross-verified. The data will be stored in the blockchain network. Sharing KYC information on the blockchain would enable financial institutions to deliver compliance outcomes, increase efficiency, and improve customer experience. KYC chain gives a solution to all the problems that occur in the existing methods. It provides a proof of identity to the customer and provides more security to the customer information.

Modules in the Blockchain-Based KYC System in Industries

- **Request to bank:** Customers can give requests to the respective bank with the KYC document, which is essential to create an account, and all the information about the customer could be stored in the KYC document. The bank staff verifies the documents and gives permission to use their account. Finally, the KYC document moves into the centralized database to create the block. Customer information can form the block as the backend of the bank. The customer requests the bank as the initial step in this module.

- **Customer view request:** As a second step, the customer views the information in the bank. The bank stored information as banking and monetary exercises straightforwardly classify the protecting stores and advances. The vast majority of these financial capacities are frequently condemned in many countries for being temperamental and helpless. Using customary monetary standards, state controllers protect the private bank stores. Those activities involving customers are utilizing this KYC chain to maintain the information as secure. So, customer can give the bank and view the details with high security by using this method. The bank gives a view request to the customer for which they have option of either allow or deny. When customer allows viewing the information, the blockchain provides a transparent view of the bank.
- **Allow or deny notification:** The third step for a customer is to permit concerned bank staff for viewing that information stored in the KYC document. All the information could be stored in the block, so all the details can be retrieved and show the customer with highly secure and low data redundancy. The arrangement of blockchain conforms to the lawful prerequisites. It also follows the information security guideline, while additionally addressing the necessities of the business and expanding its proficiency. Whose administration ensures that the square chain will fundamentally change how individual data are put away and moved among banks and customers? The primary goal of the system is to optimize payments with high security. It is not only large individual banks that study and implement blockchain. All the group of companies, the institutions, and the personnel can use this method. This high security and notification allow or deny process.
- **Transparent view:** All the information of the customer can be viewed by the concerned bank staff and the user. All the details can be listed on one page, and they can view it. Most of the financial frameworks are normally based on an incorporated database, which is increasingly helpless against digital assault in light of the fact. When the programmers assault the framework, they will get the full access over there. This revolution would dispose of a portion of the present offenses that are submitted online today against our money-related establishments. These guidelines are intended to help lessen illegal tax avoidance and psychological oppression exercises by making mandatory for organizations to confirm and recognize their customers. Blockchain would permit an association to get the subtleties of a customer checked by another association along these lines while staying away from the redundancy of the KYC procedure. The decrease in regulatory expenses for consistent divisions would be critical.

11.5 BLOCKCHAIN TECHNOLOGY FOR SECURING EHR

Health records are the utmost sensitive records associated with patients, which is recurrently communal amongst doctors, healthcare workers, and pharmacologists for effective diagnosis and treatment. For the secure storage of data, the confidentiality and sanctuary of patient records and antiquity in medical performs need to be maintained. EHR should contain the patient portal with doctor visits, lab results, and diagnosis of the diseases, which may be accessed from anywhere and anytime through peer-to-peer services. [12] Utilizing a blockchain facilitates interoperability to avoid any confidentiality and sanctuary issues. It contains intrinsic veracity and fits with legitimate guidelines. Extended interoperability gains secure information related to patient records. Much information has problem with tracking and data storing. To solve these privacy issues, smart agreements are introduced intensively. They are classified contracts, user record-associated contracts are utilized to track all patient records and maintain data integrity over the health information for secure transactions. The model utilizes explorer to manage the web-based application, composer to create smart contracts, and caliper to maintain blockchain performance over the network. Other models only have accessing rights and assigning roles to the users. There are some data integrity issues that may occur. The blockchain involves cryptographic algorithm to achieve the data integrity with extended security. The Hyperledger fabric is a private network, any transaction done between two participants is stored as well as the details between the two private miners; security is envisioned for the whole records of the patients either through transaction or risk assessments. The fabric network need not to maintain the suspended consensus rule on every node participated in the transaction. The transaction permission has been declared by the Hyperledger Burrow individually; it is a single binary distribution on following consensus rules to be accessed via the Tendermint Algorithm. For every transaction, each data is processed within the smart contracts to achieve on-chain proposal transactions.

Blockchain realizes the decentralized approach over the peer-to-peer services across various data related to patients. Data transparency can be achieved through distributed database access. The model has some properties on maintaining logs of patients and interpreted to store patient records from the hospital databases. A block contains data, previous hash code, and block hash value. Each hash code has been generated using the cryptographic algorithm. The model consists of three modules: graphical user interface layer, hybrid, blockchain layer, and smart contract layer. Each layer performs various mechanisms related to user needs.

The primary function of GUI intends to cooperate with the system and accomplish some errands such as add, update, read, etc. related to patient records. The method is based on access through a browser, which is technically known as distributed application. GUI also contains login credentials for the roles assigned to the users. The hybrid blockchain layer performs transactions (Figure 11.9) that will be needed by the users as a system function to add or delete the patient records. When the user request a patient record, it will be transferred through smart contract layer where the roles are verified for each task having some identity; in the first contract, it checks the status related to the user if it is matched, it tends to enter next arrangement where it checks the particular role is assigned for the specific patient; if it is validated, block is created or deleted by performing some consensus rules. The framework will display requested record. Blockchain plays a vital role in user validation.

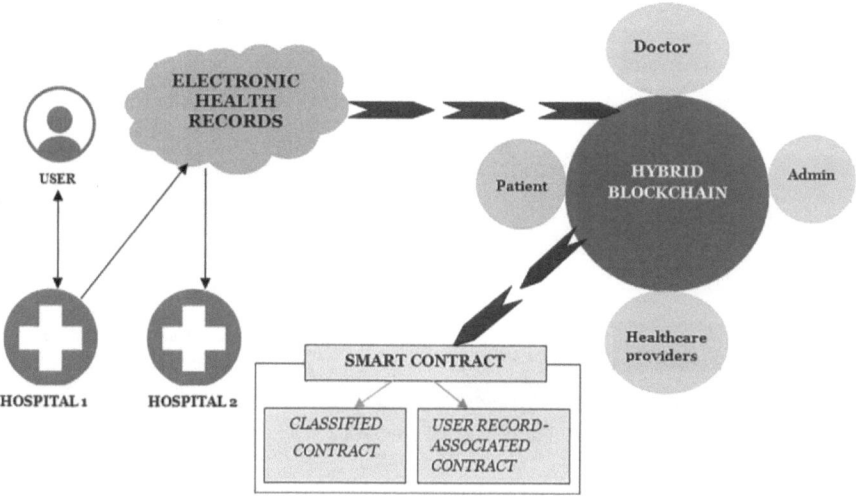

FIGURE 11.9   Blockchain architecture for EHR securing.

Blockchain is a chain of block that provides secure transaction. It is a decentralized approach. Each block is linked together in a specific order by the generated hash value; blockchains have some consensus rules to conduct system functions for validating transaction request. When the user request transaction, a block is send to each node in the network, performs block validation, and allows block into the chain. Blockchains have some key features regarding security, integrity, and anonymity to avoid unauthorized user validation. These features have made the blockchain a feasible solution for secure transactions; for the security and privacy purposes, blockchain uses SHA-256 cryptographic algorithm [13]. SHA, expanded as Secure Hashing Algorithm, is a single-way function that converts record into a 256-bits hash value for the decentralized architecture.

The smart contracts using classified contracts and user associated contracts involve validation of the doctor and the patient ID, which involves pertaining to the functional information (Figure 11.10]; while adding the block, the classified contract involves entering the doctor ID and patient ID to validate the environment of the user-associated record metadata; information involves pertaining the functional transaction to the allotted patient ID and doctor ID.

The blockchain technology uses consensus rules [14] to maintain and validate the transaction; when the user request any transaction, consensus rules need to be performed and the most important consensus rule is proof of work, which is intended to give challenges to peer networks. The peers should find the code and submit it to the proof of work. If it is valid, it will enter into the block; otherwise, the block is denied.

Smart contracts are known as the chain code which is used to execute some rules of the system function. A smart contract tends to maintain the required operations in a scalable framework; when the user requests transaction, it will execute the section of code. It is the most important agreement in the blockchain technology.

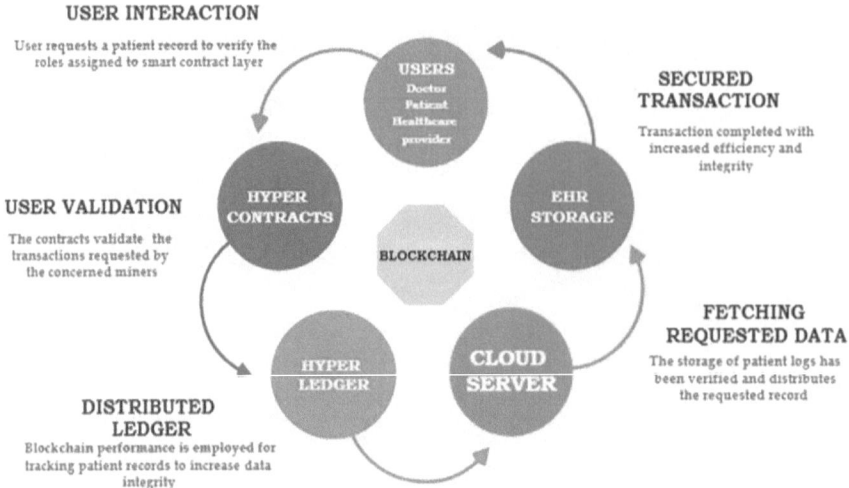

FIGURE 11.10   Blockchain process using smart contracts.

Hyperledger is an open-source project used to make the blockchain-based relations. Hyperledger have some tools to explore the health records sensitively:

- Hyperledger explorer involves creating a web-based application on evolved environment.
- Hyperledger composer is used to build a business networks.
- Hyperledger caliper is intended to ration the enactment of the blockchain network.
- Hyperledger Ursa involves maintaining and ignoring the hashing functions.

11.6   BLOCKCHAIN-BASED FOOD SUPPLY CHAIN MANAGEMENT

Generally, blockchain is based on three techniques: public, private, and consortium. Open model is transparent to every user in the network. If any block in the network updates its transaction, it means it will be visible to every other in the system and at any time the new block will able to join the network. Because of its public transparency, the model is said to be permission-less model; if a new block is requested to join in the network, it means it will join quickly without any permission. An another technique is based on a private model where new block requires permission from a single owner, and the network updates its node if and only if it gets permission from an owner. The third technique is the consortium model, which is also similar to the private model, and the difference is it contains several owners instead of the single owner. Private and consortium models are said to be permissioned blockchain. A comparison of different blockchain models is given in Table 11.1.

Food supply chain model is a complex process where the quality of food is very important. In order to ensure food safety, it is required to evaluate the food credits by

TABLE 11.1

Comparison of Blockchain Models

Permission-Less Blockchain	Permissioned Blockchain
The node does not need any permission, each node should take part	The node requires permission from any central authority to perform certain operations
High latency	Higher efficiency
It is fully decentralized	It is partially centralized

the traders [4]. The supply chain begins from food providers (farmers) to consumers through multiple stakeholders like factory productions, retailers, and distributors. The traders have to ensure the food safety to the consumer, but traders randomly increase their trade point without any supervision; it leads to multiple food scandals like dog meat scandal in India in 2019 and horse meat scandal in the UK in 2013. And today the COVID-19 pandemic is because of the pangolin meat in wet market of Wuhan, China, because of lack of supervision in food supply chain and also because traders provide information selectively in order to increase their profit. So, the food supply chain requires proper supervision and credit evolution. Recent studies report many innovations of blockchain in food industries. A combination of deep learning and blockchain technology will help to evaluate the credit and supervise the food supply system.

Food supply system contains traders and supervisors. Both are connected through the blockchain network, and their transaction is updated in the network based on permissioned blockchain. The credits are evaluated by the deep learning model called Gated Recurrent Unit (GRU).

Traders are the different types of stakeholders involved in the food supply chain: farmers, consumers, retailers, etc. In permissioned blockchain, they are represented through the individual block in a peer-to-peer network. Traders are responsible for credit evolution and updated stock exchange in the system.

Supervisors are responsible for monitoring the traders at a regular interval and also maintain the food chain supply. They have an authority to manage the traders and maintain the transaction of traders in the network.

GRU is a special type of Recurrent Neural Network (RNN), and it is based on a discriminative deep learning model. GRU is considered as an advancement of the LSTM model, and it requires minimum computation time compared to LSTM with higher accuracy. In food supply chain, GRU is used to analyze the credit by using stakeholder's transactions done through blockchain. The workflow of credit evolution system is illustrated in Figure 11.11.

- Trader X and trader Y are two blocks in the blockchain network. Both traders complete their transaction, i.e., exchange of crops from trader X to trader Y.
- Trader Y replies a credit evolution to the trader X based on the product quality and trader satisfaction; for example, product is good.

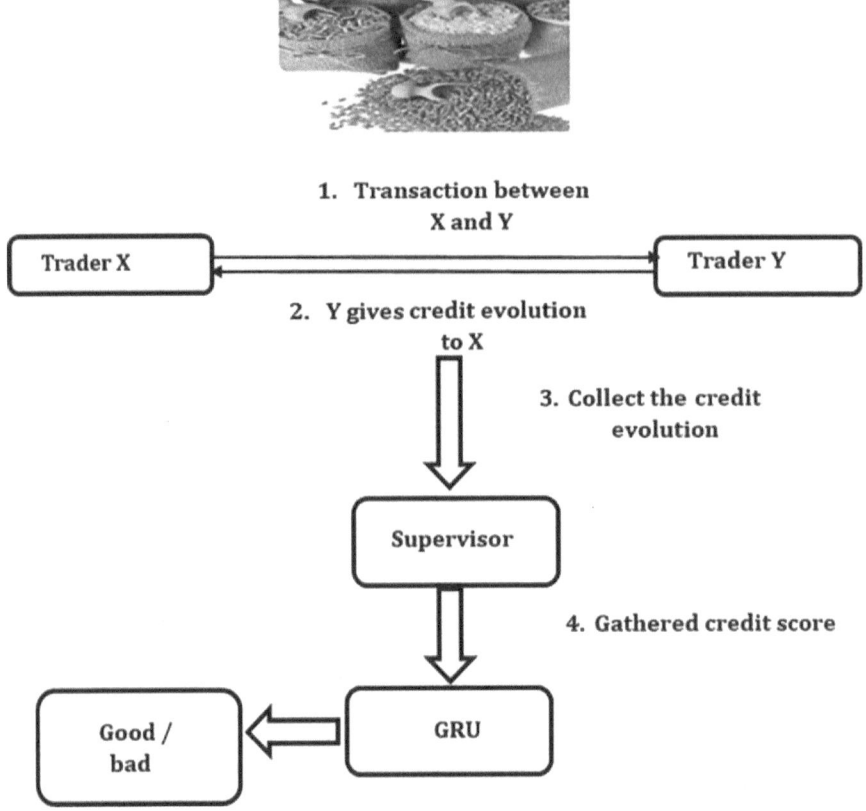

FIGURE 11.11   Working of food supply chain.

- Supervisor collects the credit evolution of various nodes in the network like Z Y A B at a regular interval of time (e.g. 1 week, 1 month).
- The gathered credit evolution texts can now be trained by using binary GRU deep learning models to analyze the trader's credits based on the texts.
- Finally, the GRU model output the texts based on the keywords. If it finds any keywords similar to a good, then it results in positive feedback or else it results in negative feedback. In our example, the result is positive.

11.7   BLOCKCHAIN APPLICATIONS IN DIGITAL INDIA

Blockchain is an emerging technology in today's world that replaces the idea of centralized server. In our day-to-day lives, everything depends on online. As a result, it produces millions of data, and these data are collected and maintained through a centralized server. The main issue faced by the centralized server is lack of security, computing failure, etc. Here, revolutionary technology blockchain has proved its impact in various fields with the help of the decentralized server and its transparency nature. Initially, blockchain was used by the cryptocurrency like bitcoin, Litecoin, etc. Blockchain is now being used in the various application fields like agriculture and healthcare in combination with emerging technologies like IoT, artificial intelligence, and cloud computing.

Aritra Brahma and Dutta [15] planned a cashless exchange framework where the transactions are managed without utilizing the real money or the methods for hard cash. The RBI and the Government of India are making attempts to lessen the utilization of funds in the economy by advancing the computerized installment gadgets, including prepaid instruments and cards. RBI's push to support these new assortments of installment and repayment offices intends to accomplish the objective of a less money society. This work centers on the benefits, possibilities, and security dangers, identifying with the cashless exchanges.

Naveen Kumar [16] has shown how the financial area has assumed a significant position in monetary improvement and development, because starting with one source and then moving onto the next, the assets transfer from savers to speculators, thereby helping the nation in its work. A sound money-related framework in the United States is quickening the pace of the capital and financial growth and assisting with accomplishing development by giving satisfactory assets and credits to different areas and motivating individuals to keep money to the very much earned. India's financial framework works and sets up a business by giving credit and advances to different areas, utilizing a lot of open private and outside bank assets to create accessible and corporate assets.

Sven Kiljan et al. [17] portrayed the significance of confirmation in the data security field. Although much expounded on surveying substance (client) verification strategies, there is yet an absence of writing concerning the assessment of money-related exchange validation in web-based banking. Substance verification strategies have been systematized by measuring their subjective viewpoints. However, no assessment component adds to the extra attributes of exchange validation in a client-driven setting. Because of a current instrument that measures openness, memorability, security, and weakness qualities in element verification techniques, we propose possibility as a new measurement that evaluates viewpoints identified with the safe convenience of exchange confirmation strategies. The utilization of this assessment instrument is proposed by numerous raters to lessen individual predisposition. Four executives and seven specialists assessed eight proposed confirmation techniques for web-based banking. The outcomes demonstrate that the instrument can be applied on a broad scope of verification techniques since it can determine strategies dependent on various data plans. In any case, care must be taken that assessments are performed by numerous specialists, because of the measure of subjectivity characteristic in the system and the various evaluations of the raters.

Brian Soeder and Barber [18] explored why online exchanges require an essential connection among clients and asset suppliers (e.g., retailers, banks, Internet-based life systems) based on trust; two clients and suppliers must accept the individual or association they metal interfacing. However, as time passes, significant information ruptures and other personality-related cybercrimes become an everyday lifestyle, and existing strategies for client character confirmation are not sufficient. Besides, much research on personality reliability centers on the client's point of view, while asset suppliers get less consideration. Along these lines, about it is being researched how suppliers can improve the probability their clients' personalities are dependable. Based on the existing research, the client–supplier trust relationship can be displayed with various transaction settings and characteristics of nature. The model was investigated for two parts of client's personality: dependability with unwavering quality and validity with a critical arrangement of real client characters obtained from the U.S. Branch of Homeland Security. In general, this research finds that asset suppliers can essentially build trust in client character reliability by just gathering a constrained measure of client personality qualities.

Ritu and Renu [19] researched the expansion of remote gadgets, offering availability and comfort; they keep on applying sublime requests on vendors to send secure remote-based electronic business applications. The critical feature, which would associate with maintaining this prerequisite, is a significant opportunity for supporting both electronic charge card and plastic exchanges over remote systems. Even though the Secure Electronic Transaction (SET) convention executes a start-to-finish security for Master Card exchanges over a different framework, a few elements, including transmission capacity prerequisites, accompany unsatisfactory remote applications.

Azaria et al. [20] recommended a decentralized record executive framework to deal with EHRs. The framework gives patients a permanent log and simple access to their restorative data. Maesa et al. [21] suggested a blockchain-based answer for distributing strategies for communicating to get privileges of assets and permitting rights among clients, the policies, and the rights have been exposed in a distributed form. The creators displayed a bitcoin-based confirmation idea usage without any investigation or assessment. Wang et al. [22] proposed a mobile application called HDG based on blockchain consensus for patients to enable them to share medical records with indicator-centric schema that has been used to organize the patient's data in an easy way to avoid the intermediary interventions. Xia et al. [23] proposed a patient-driven framework using a cryptographic algorithm to restrict unauthorized users from sharing patient data between the healthcare and the cloud service custodians to improve data privacy while distributing medical records. It is the most prominent way to access the records and maintain them.

11.8 CONCLUSION

Blockchain applications go a long way past digital money and bitcoin. With its capacity to make more straightforwardness and decency while likewise setting aside organizations' time and cash, the innovation is affecting an assortment of divisions in manners that run from how agreements are upheld to making government work all the more effectively. Distributed ledger and Ethereum innovation holds colossal potential for improving effectiveness, giving continuous data, and diminishing expenses. Blockchain can make patients in charge of their information by empowering them with access to EHRs through a fine-grained consent framework. It can give an advanced mark to every patient data utilizing cryptographic technologies. Also, blockchain serves a real need by diminishing KYC regulatory expenses and lost time while simultaneously expanding security and straightforwardness. Finally, blockchain technology coupled with the ability to program business logic with the use of smart contracts enables the provenance of consumer goods from the source point to end consumption with accurate asset tracking. Based on our study, we conclude that blockchain is a promising technology for next-generation product customization and client services in all sectors of Industry 4.0.

REFERENCES

1. M. H. Miraz and M. Ali, "Applications of Blockchain Technology Beyond Cryptocurrency," *Annals of Emerging Technologies in Computing (AETiC)*, vol. 2, no. 1, pp. 1–6, 1st January 2018.
2. Z. Shae and J. Tsai, "Transform Blockchain into Distributed Parallel Computing Architecture for Precision Medicine," *IEEE 38th International Conference on Distributed Computing Systems*, (ICDCS) 2018. (pp. 1290–1299).
3. J. Parra Moyano and O. Ross, "KYC Optimization Using Distributed Ledger Technology," *Business & Information Systems Engineering*, vol. 59, pp. 411–423, 2017. https://doi.org/10.1007/s12599-017-0504-2
4. F. Tian, "An Agrifood Supply Chain Traceability System for China Based on RFID & Blockchain Technology,". In *2016 13th international conference on service systems and service management (ICSSSM)*, pp. 1–6. IEEE, 2016.
5. C. Holotescu, "Understanding Blockchain Technology and How to Get Involved," *The 14th International Scientific Conference eLearning and Software for Education Bucharest*, April 2018. pp. 19–20.
6. P. K. Sharma and J. Hyuk Park, "Blockchain-Based Hybrid Network Architecture for the Smart City," *Future Generation Computer Systems*, vol. 86, pp. 650–655, 2018.
7. A. Brahma and R. Dutt, "Cashless Transactions and Its Impact - A Wise Move towards Digital India," *International Journal of Scientific Research in Computer Science, Engineering and Information Technology*, vol. 3, no. 3, pp. 14–28, March 2018.
8. K. Bhaskaran et al., "Double-Blind Consent-Driven Data Sharing on Blockchain," *2018 IEEE International Conference on Cloud Engineering (IC2E)*, 2018, pp. 385–391.
9. W. M. Shbair, M. Steichen, J. François and R. State, "Blockchain Orchestration and Experimentation Framework: A Case Study of KYC," *NOMS 2018–2018 IEEE/IFIP Network Operations and Management Symposium*, Taipei, 2018, pp. 1–6.
10. P. C. Mondal, R. Deb and M. N. Huda, "Transaction Authorization from Know Your Customer (KYC) Information in Online Banking," *2016 9th International Conference on Electrical and Computer Engineering (ICECE)*, Dhaka, 2016, pp. 523–526.
11. W. Cai, Z. Wang, J. B. Ernst, Z. Hong, C. Feng and V. C. M. Leung, "Decentralized Applications: The Blockchain-Empowered Software System," in *IEEE Access*, vol. 6, pp. 53019–53033, 2018.
12. G. Kamau, C. Boore, E. Maina and S. Njenga, "Blockchain Technology: Is This the Solution to EMR Interoperability and Security Issues in Developing Countries?" in *2018 IST-Africa Week Conference (IST-Africa)*, IEEE, 2018, pp. 1–8.
13. Q. Xia, E. B. Sifah, K. O. Asamoah, J. Gao, X. Du and M. Guizani, "Medshare: Trust-Less Medical Data Sharing among Cloud Service Providers via Blockchain," *IEEE Access*, vol. 5, pp. 14 757–14 767, 2017.

14. Y. Ji, J. Zhang, J. Ma, C. Yang and X. Yao, "BMPLS: Blockchain Based Multi-Level Privacy-Preserving Location Sharing Scheme for Telecare Medical Information Systems," *Journal of Medical Systems*, vol. 42, p. 147, 2018.

15. A. Brahma and R. Dutta, "Cashless Transactions and Its Impact - A Wise Move towards Digital India," *International Journal of Scientific Research in Computer Science, Engineering and Information Technology (IJSRCSEIT)*, vol. 3, no. 3, pp. 14–28, March-April–2018.

16. Naveen P. G. Kumar, "Use of Information Technology in Banking Transactions Using Electronic Payment Systems," *Journal of Bank Management & Financial Strategies*, vol. 1, no. 2, pp. 33–42, August, 2017.

17. S. Kiljan, H. Vranken and M. van Eekelen, "Evaluation of Transaction Authentication Methods for Online Banking," *Future Generation Computer Systems*, vol. 80, pp. 430–447, 2018.

18. B. Soeder and K. S. Barber, "A Model for Calculating User-Identity Trustworthiness in Online Transactions," *2015 13th Annual Conference on Privacy, Security and Trust (PST)*, Izmir, 2015, pp. 177–185.

19. Ritu and Renu, "Research on Online Transaction Protocols for Supporting Credit/Debit Card Transaction," 2014.

20. A. Azaria, A. Ekblaw, T. Vieira, and A. Lippman, "Medrec: Using Blockchain for Medical Data Access and Permission Management," in *2016 2nd International Conference on Open and Big Data (OBD)*, IEEE, 2016, pp. 25–30.

21. D. D. F. Maesa, P. Mori and L. Ricci, "Blockchain Based Access Control," in *IFIP International Conference on Distributed Applications and Interoperable Systems*, Springer, 2017, pp. 206–220.

22. X. Yue, H. Wang, D. Jin, M. Li and W. Jiang, "Healthcare Data Gateways: Found Healthcare Intelligence on Blockchain with Novel Privacy Risk Control," *Journal of Medical Systems*, vol. 40, no. 10, p. 218 (8 pages), 2016.

23. Q. Xia, E. B. Sifah, K. O. Asamoah, J. Gao, X. Du and M. Guizani, "Medshare: Trust-Less Medical Data Sharing among Cloud Service Providers via Blockchain," *IEEE Access*, vol. 5, pp. 14 757–14 767, 2017.

12 The Winning Combo
Cryptocurrency and Blockchain

S. Matilda and T. Ananth Kumar
IFET College of Engineering

CONTENTS

12.1 INTRODUCTION

The world that we live in revolves around money and is witnessing a cryptocurrency movement. With the world economies fluctuating rapidly, financial services of different types have emerged to support organizations and deal with their finances and other resources. These financial institutions are designed to support economies of developing and developed countries with varied income ranges. Radical change can be observed in the manner in which these financial institutions and transactions have matured over the years. Over generations the modes of financial transactions have taken different façades – starting from exchange of goods to minting of copper coins to currency system where transactions were held face-to-face. With the advent of Internet, things changed overnight and everything went online. The people involved were known, but their physical presence was not necessary. The new avatar is cryptocurrency where transactions are transparent, but everyone involved remains anonymous. It presents a new perspective of monetary exchange and revolves around blockchain, miners, cryptography, and wallets. It allows peer-to-peer electronic cash transaction that has cryptographic proof instead of trust. The primary characteristics based on which cryptocurrency functions are trust, immutability, and decentralization. As of July 2020, there are around 5,784 cryptocurrencies [1]. Bitcoin, the one for which blockchain technology was invented, rules the market with a current dominance of about 61.4% [1].

12.2 A LITTLE OF HISTORY: THE ASCENT OF MONEY

The kings of the ancient world knew that "Money is power" as it gave them absolute monopoly. That is the reason why they minted copper coins and asked people to buy things and pay their taxes using coins. At the point when realms were rearranged as country expresses, the ability to print cash went to the state and any individual who attempted to make their own cash got squashed. The explanation is one incorporated adversary can without much of a stretch be wrecked with a "beheading assault" according to the platitude "Remove the leader of the snake and that is the end". Along these lines nobody would set out test the intensity of the single position and its heavenly option to make coins.

This was seen when e-gold was acquainted in an endeavor to make an elective money. It had over a million records at its top in 2008, and was handling over $2 billion dollars of exchanges. The U.S. government assaulted the four heads of the framework, bringing charges against them for running an "unlicensed cash transmitting" business. It obliterated the organization by bankrupting the authors and the game was finished. "Unlicensed" is the catchphrase in the assault and the store expression is "Control the cash and you control the world".

In decentralized frameworks, there is no leader of the snake. Decentralized frameworks are a hydra: "Cut off one head and two more would fly in to have its spot" [2]. That is the motivation behind why Satoshi, the originator of bitcoin, the main decentralized digital money, carefully stayed mysterious. He named bitcoin as a "Distributed Electronic Cash System".

12.3   MONEY AND ITS PROPERTIES

Assume that you like to credit a few dollars in your bank account. Let's say that you take one of the following items, such as your homegrown organic vegetables, or a brand new Kashmiri carpet, or your child's favorite toy and try depositing them in a bank for an equivalent value. The bank would not accept this as it does not fall under the basic description of money. The vegetables are perishable and its value depreciates with time. Economists call this as "store of value". If you want to buy any item that costs one-fourth the price of the Kashmiri carpet, you cannot exchange it with a quarter piece of the carpet. This concept is called "unit of account". Your child's favorite toy is valuable to you and your child, but not to anybody else. Nobody would exchange it for any goods or service, as it does not assist as a "standard medium of exchange". But a dollar bill is accepted by one and all, as it satisfies the aforementioned qualities. The dollar bill is accepted based on the trust that its value today will hold good for tomorrow.

Cash is further characterized as anything or accurate record that is commonly adequate by the individuals in return of products and ventures or reimbursement of obligations. Customarily, cash plays out the accompanying four principal works, the essential being "mechanism of trade" and "proportion of significant worth". "Standard of conceded installment" and "store of significant worth" are called optional capacities since they are gotten from the essential capacities. Another capacity that business analysts stress nowadays is "Liquidity of Money", which empowers convertibility into money. The capacity to change over an advantage into cash rapidly without loss of significant worth is called liquidity of benefit.

Financial specialists separate cash into three distinct sorts: item cash, fiat cash, and bank cash. Item cash is a decent whose worth fills in as the estimation of cash [3]. Gold coins are a case of item cash. Advanced cash is a fiat money, which implies that the paper on which it is printed is useless. Worth expansion is a direct result of the promissory note that says "I guarantee to pay the conveyor an entirety of" marked by the Reserve Bank Governor of the nation; for example, Indian rupee, dollar note. Bank cash comprises of the book credit that banks stretch out to their contributors. Exchanges are made utilizing checks, net banking, or credit and charge cards on the store held by the individual [3].

12.4   TYPES OF CURRENCIES

Money in any form typically bank notes and coins, which are used in circulation as a medium of exchange are referred to as "currency" [4,5]. In other words, it is "a system of money which is commonly used by the people of a nation". For example, Indian rupee (INR), yen (¥), U.S. dollar (USD/US$), euro (€), and pound sterling (£) are some currencies that are used across the globe. These currencies are used for trading with other countries in foreign exchange markets. The magnitude of this trading has partial limitations of acceptance, as defined by the relative value of the currency.

Digital currency is a generic term for all electronic money, and it encompasses regulated and unregulated, virtual currency and cryptocurrency. They are used in electronic wallets or designated transaction networks. Entire virtual currencies are meant to be digital, but *not* all digital currencies are meant to be virtual. Virtual currencies are controlled by its originators and are usually used for transactions amongst the associates of a closed virtual community for peer-to-peer payments [6]. Most virtual currencies are in the form of tokens and are subjected to heavy price swings as they are unregulated.

Another form of currency is cryptocurrency. The word "crypto" conjectures that cryptographic techniques and encryption algorithms ensure security in the transactions performed. The probability of making a counterfeit cryptocurrency is extremely low, as they operate as blockchain-based decentralized systems without the necessity for a trusted third party like a financial institution or MasterCard company. During this instance, peer-to-peer transfers are smoothed through the exploitation of personal as well as public keys. Currently, bitcoin is the most well-known and widely used blockchain-based cryptocurrency. There are numerous other options, or altcoins; for example, Ethereum, XRP, Tether, Litecoin, Tezos, and Monero [1]. A portion of these recreates bitcoin innovation, while others are slight variations or new cryptographic forms of money that have been partially gotten from existing ones.

Cryptographic forms of money such as bitcoin and Ethereum are arranged as virtual monetary forms. This infers as opposed to advanced cash; it conveys some financial incentive in electronic structure; however, it stays outside the domain of any administrative body. Because of their virtual nature, cryptographic forms of money don't have a focal storehouse, which shows that they can be crushed by a PC crash if there is no reinforcement duplicate of the property or if the client loses their private key. Dissimilar to mysterious money exchanges, exchanges completed with digital forms of money can be followed on the blockchain without at first knowing the personality of the issuer. Some cryptographic forms of money such as Dash, Zcash, and Monero are less private than others and are unquestionably more difficult to follow than bitcoin. Value unpredictability is an unmistakable character of cryptographic money as its worth depends entirely on market interest.

12.5 FEATURES OF CRYPTOCURRENCY

The framework that defines any cryptocurrency sets it apart from the rest. Any cryptocurrency can be identified by the prominent key features, which are described in the following sections.

12.5.1 DECENTRALIZATION

The concept of decentralization is to take power off from one source and sharing it with everyone involved. The true influence of cryptocurrencies is the authority to create and distribute money without a central power. In traditional fiat currencies, there is a central authority, such as banks, that controls all the financial transactions for which a sizable amount is to be paid as fees. However, with cryptocurrencies, using distributed and open networks, these transactions are processed and validated in which everyone is a part. Hence, transaction costs are close to nothing. Every person in possession of a cryptocurrency is called as a node and all the transactions performed by the nodes are updated and recorded in a public distributed ledger. These data are encrypted and can be accessed and verified by all the nodes. The transaction is proliferated across the entire network, replicated by every node, and reaches a large number of nodes within a few seconds. Thus, there is no single node that has the authority to control the network.

12.5.2 IRREVERSIBLE

The irreversible and unchanging component of digital currency suggests that exchanges can't be changed when it is recorded on the appropriated record. It is unthinkable for anybody, yet the hub has the separate private key to move its excellent resources. To forestall deceitful exchanges (that can't be turned around), all exchanges are straightforwardly recorded on the blockchain and open to general society. While it isn't challenging to alter the exchange, secure cryptography makes it extremely hard for alteration, since it expects you to bargain all the hubs in the blockchain. When the exchange is affirmed, nobody can support you, on the off chance that you have sent assets to a con artist or made an off-base exchange.

12.5.3 PSEUDONYMOUS

Pseudonymity is a term coined from the words pseudo and anonymity. Hence, pseudonymity exhibits the anonymity property in the sense that anyone can purchase or possess a cryptocurrency without revealing their real identity. For example, bitcoin uses public key hashes as address that acts as pseudo-identities and do not reveal the real identities. But this can be broken and the identity revealed if an adversary is able to create a pattern of the transactions made by a person. Hence, pseudonymity also mirrors the unlikability property, in the context that different transactions of the same user cannot be linkable to each other. In a more clear sense, when a client associates with the framework over and again, these various collaborations ought not have the option to be attached to one another from the perspective of a foe.

12.5.4 FAST AND GLOBAL

Transactions are communicated in a split second to the whole system and are avowed inside a couple of moments. Since they occur in a universal system of PCs, they are totally aloof of the physical area. Henceforth, the speed of exchange is simply the speed of the Internet. There are no firm principles for purchasing cryptographic forms of money. Anybody living in any piece of the globe can turn into an individual from the system.

12.5.5 SECURITY

Cryptocurrencies are built on cryptography and appear as entries in distributed consensus databases. They are called CRYPTO-monetary standards on the grounds that the accord-keeping process is made ensured by solid cryptography. They are not ensured by entities or by trust, but by math [7]. The probability of a cryptocurrency address being compromised is very remote.

12.6 TYPES OF CRYPTOCURRENCY

12.6.1 BITCOIN

The first and most popular blockchain-based cryptocurrency is bitcoin. It is the most valuable cryptocurrency. Bitcoin is a combination of several novel technologies (cryptography, peer-to-peer networks, distributed databases). The unit of account of the bitcoin system is a bitcoin. Symbols used to represent bitcoin are BTC and XBT. Small amounts of bitcoin are millibitcoin (mBTC) and satoshi (sat). In 2009, bitcoin was launched in 2009 with a total market value of around $157 billion by Satoshi Nakamoto [8]. Just as U.S. dollar is divisible to two decimals, bitcoin is divisible down to eight decimals. Bitcoin has an anonymous nature that is imparted by the crypto-algorithms and at the same time is transparent. The transparency is brought about by the visibility of the structure of the transactions and the coding that anyone in the network can use to build the blocks.

Today, there are multiple alternatives for cryptocurrencies with various functions and specifications. A number of these are clones or forks of bitcoin, while others are new currencies that were built from scratch. A number of competing cryptocurrencies spawned by bitcoin's success, referred to as "altcoins", include Litecoin, Peercoin, and Namecoin, and also as Ethereum, Cardano, and EOS. Today, the mixture value of all the cryptocurrencies alive is around $214 billion—bitcoin currently represents about 68% of the entire value [9].

12.6.2 ETHEREUM

In 2015, Ethereum is launched as the leading programmable blockchain and its cryptocurrency is called Ether (ETH). ETH replicates most features of bitcoin and is purely digital. Transactions are instant and are generally used to make payments, as a store of value, or as collateral. Unlike other blockchains, Ethereum is programmable, which means that developers can use it to build new kinds of applications.

Some of the financial applications of Ethereum are as follows:

- Cryptocurrency wallets that aids in instant payments with ETH
- Financial submissions that serve to borrow, lend, or invest in digital assets
- Decentralized marketplaces, which facilitate trade of digital assets and predictions about events in the real world [10].

12.6.3 LITECOIN

Litecoin is the second most well-known cryptographic money. Litecoin is one of the shared Internet cash that empowers close to zero cost installments to all people. Litecoin is an open-source, worldwide repayment organizer that is entirely decentralized, and the security highlights enable people to control their own financial records. Contrasted with bitcoin, Litecoin performs quicker exchange (normal affirmation time of 2.5 minutes) and improved stockpiling proficiency. With significant industry support, exchange volume, and fluidity, Litecoin is also a demonstrated mode of business integral to bitcoin [11].

12.6.4 ZCASH

Zcash permits straightforward exchanges to oblige for wallets and trades that don't bolster private exchanges. Zcash addresses are either Z-addresses (private) or T-addresses (straightforward). T-addresses start with a "t" and Z-addresses start with a "z". The z-to-z exchange shows up on the open blockchain; so it is recognized to have happened and that the expenditure has been paid. In any case, the addresses, exchange sum, and the update field are scrambled and not openly unmistakable. The proprietor of a location may decide to reveal Z-address and exchange subtleties and confided in outsiders using view keys and installment exposure.

Exchanges between two straightforward locations (T-addresses) work simply like bitcoin: The sender, beneficiary, and exchange esteem are openly noticeable. While numerous wallets and trades solely use T-addresses, many are moving to protected locations for better client security. The two Zcash address types are always interoperable. But the assets can be moved in-between T-addresses and Z-locations. Zcash is endorsed by the community and the most important exchanges. Zcash can be shifted from Zcash to other cryptocurrencies and into cash, whereas many off-blockchain systems do not easily allow people to withdraw and redeem it in the real world.

12.6.5 RIPPLE

Ripple empowers banks, installment suppliers, computerized resource trades, and corporates to send cash all-inclusive, utilizing progressed blockchain innovation. Ripple Net offers the most developed blockchain innovation for worldwide install-ments, which makes it simple for budgetary establishments to arrive at a believed, developing system of more than 300 given around 40 nations and six landmasses. The group at Ripple would like to empower the world to move esteem, like it as of now moves data on the web today [1].

12.6.6 TETHER

Tether is questionable cryptographic money with tokens given by Tether Limited. Tether Limited and the digital tether currency are dubious, given the organization's inability to give a guaranteed review demonstrating sufficient stores backing tie, and it is supposed job in controlling the cost of bitcoin. The tether was initially intended to be a stable coin in light of the fact that its worth would consistently be $1.00. On 30 April 2019, Tether Limited's attorney guaranteed that each tether was upheld by just $0.74 in real money and money counterparts [12–14].

12.7   TRANSACTION PROCESS USING BLOCKCHAIN

Cryptocurrency is one of a few hundred applications that utilizes blockchain innovation. Prior blockchain applications required a perplexing coding, cryptography, and science just as noteworthy assets. Lately, devices have been created to fabricate decentralized applications, for example, electronic democratic, carefully recorded property resources, administrative consistency, and exchanging. Every one of these applications is created and conveyed quicker than at any other time. The electronic coin may be a chain of computerized marks. Every proprietor moves the coin to somebody else by carefully marking a hash of the past exchange and therefore the open key of the subsequent proprietor and adding these as far as possible to the coin. A payee can check the marks to verify the chain of possession. This situation is reflected in Figure 12.1.

The digital money industry makes it workable for people to perform exchanges over a protected stage without uncovering their characters. It is a true innovation that is decentralized naturally and means to empower, assent, and register digital currency moves. Blockchain innovation extraordinarily disentangles the immediate exchange of exchange resources and builds trust in the members of the exchange. This is accomplished by giving one of a kind, nonforgeable characters for resources, alongside a sacred record of the proprietorship.

Digital ledger is the utensil behind blockchain, which plays out the budgetary exchanges, by recording all the exchanges that can't be undermined. The record

FIGURE 12.1   Digital signature.

makes a reasonable, decentralized record of exchanges that permits the substitution of a solitary ace database. It keeps a changeless record, everything being equal, back to the beginning purpose of an exchange. This is otherwise called the provenance, which is fundamental in the exchange account, permitting money-related establishments to audit all exchange steps and decrease the danger of extortion.

A blockchain, from the digital currency point of view, is often characterized as a period-stepped arrangement of permanent records of data that's overseen by a bunch of PCs not possessed by any single substance. All of those exchanges (for example, square) are formed, made sure about and sure to each other utilizing cryptographic standards (for example, chain). A blockchain conveys no exchange cost, although it features a microscopic framework cost. The blockchain may be a primary yet astute method for passing data from shared in a wholly robotized and safe way. One gathering of exchange starts the procedure by making a square. Thousands check this square, maybe an enormous number of PCs circulated round the net. The checked square is added to a sequence, which is put away over Internet, making an unprecedented record, yet one of a sorting records with a unique history. Misrepresenting a single record would mean adulterating the entire chain during a great many occasions, which features a distant chance. Bitcoin utilizes this model for money-related exchanges. However, it could also be sent from numerous points of view. Blockchain is out to supplant every incorporated procedure and plans of action that make sudden spike in demand for commissions charged for exchanges.

Blockchain empowers move of computerized coins starting with one individual and then onto the next. On the off chance that Alice needs to move cash to Bob, from India to Japan, this is normally done utilizing a third confided in party, ordinarily much of the time money-related organizations like banks. It fills in as follows: Alice says, " I need to move cash to Bob in London" and gives the duty to the confided outsider. The believed outsider distinguishes Bob in London and moves the cash to Bob's record subsequent to gathering a charge for the administration, as appeared in Figure 12.2.

Blockchain attempts to solve this without the trusted entity so that Alice and Bob can talk with each other directly and do the transfer immediately. In addition, it is cheaper than the fee that the third party gathers. The basic principle behind which this works is the concept of "open ledger". This can be illustrated using a network of four people: Alice, Bob, Charles, and Diana, who want to move money from one to another. Let us assume that at the moment of the inception of this network, Alice has 100 dollars.

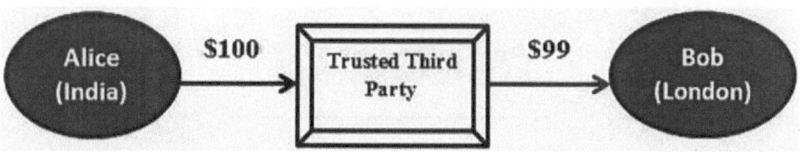

FIGURE 12.2 Transaction using third party.

12.7.1 Open Ledger

If Alice wants to move 50 dollars to Bob, the transaction is effected and is linked to the already existing transactions. A transaction is also a file that says, "Alice gives $50 to Bob" and is signed by Alice's private key. It is basic public key cryptographic technique. If Bob wants to move to Diana $30, the transaction is effected and is linked to the previous transaction. Now these transactions and links form the ledger. The same procedure is followed if Diana wants to transfer $10 to Charles. As the transactions are visible to everyone in the network, it is said to be an open ledger. Figure 12.3 shows the structure of an open ledger that is essentially a chain of transactions. For this reason, it is called as a blockchain. It is a chain of transactions that is open and public to everyone. Everyone owns the network and can see where the money is, how much money each one has, and everyone can decide whether a transaction is valid or not.

For example, let's say that Alice now attempts to move $80 to Charles. As everyone owns the network, they can immediately see that this is not a valid transaction as Alice has only $50 in her account. This transaction will not be added to the open ledger and will not be part of the chain. In this case, the ledger is centralized with everyone in the network having access to it. But the goal of blockchain is decentralization, which works only when we have a distributed ledger.

12.7.2 Distributed Ledger

The concept is to distribute the ledger across various nodes in the network, and everyone in the network can see these ledgers as they are open. The nodes thus have a record of the complete history of all transactions and the balance of every account. Figure 12.4 demonstrates a distributed ledger where Alice and Diana have a copy of

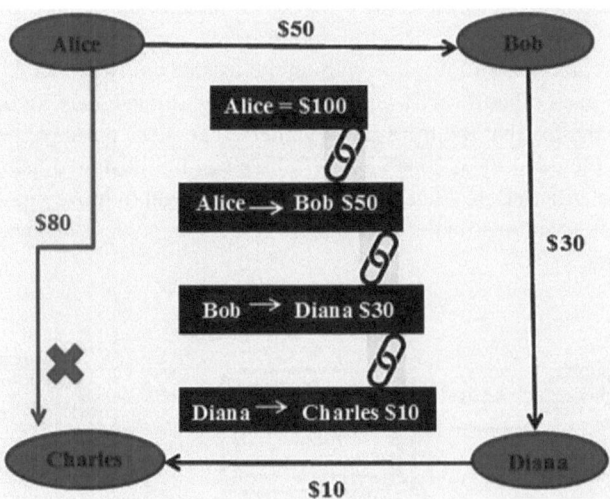

FIGURE 12.3 Open ledger.

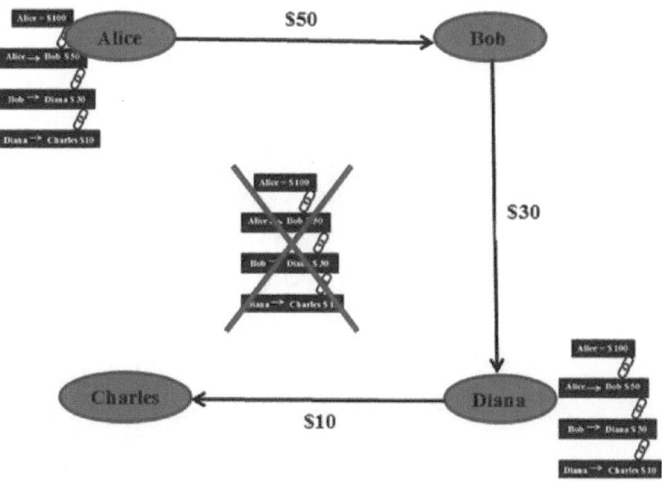

FIGURE 12.4 Distributed ledger.

the ledger. There is no restriction as to who can have a copy of the ledger as everyone can see the ledger. This means that all the nodes, Alice, Bob, Charles, Diana, and anyone else who are part of the network, will either have a copy of the ledger or can have access to the ledger. Once this is done, we can do away with the centralized ledger as it is no longer required.

But this brings in a new problem and leads to the third principle of blockchain, which is probably the most interesting one. When there are multiple copies of the ledger in the network, we need to ensure that all these copies are synchronized and that all the participants in the network see the same copy and the same version of the ledger. Consider the transaction where Bob wants to transfer $30 to Diana. Once this transaction is intended, Bob broadcasts the intention as a message to everyone in the network. It uses the basic p2p technology, and message is sent from one peer to every other peer. This is still an invalidated transaction and is not entered in the ledger. Confirmation is a critical concept in cryptocurrencies. As long as a transaction is not validated, it is in an unconfirmed state. As confirmation is still pending, the transaction can be forged. When a transaction is confirmed, it becomes part of an immutable record of historical transactions, also called blockchain. It is no longer forgeable and cannot be reversed.

12.7.3 CRYPTOCURRENCY MINING

To comprehend the system of affirming the exchange and making a passage in the record, we have to comprehend the idea of diggers. Excavators are extraordinary hubs that hold the record. Basically, everyone can be a digger. Since a decentralized system has no position to appoint this errand, a component is required to keep one hub from manhandling or commanding another. On the off chance that somebody

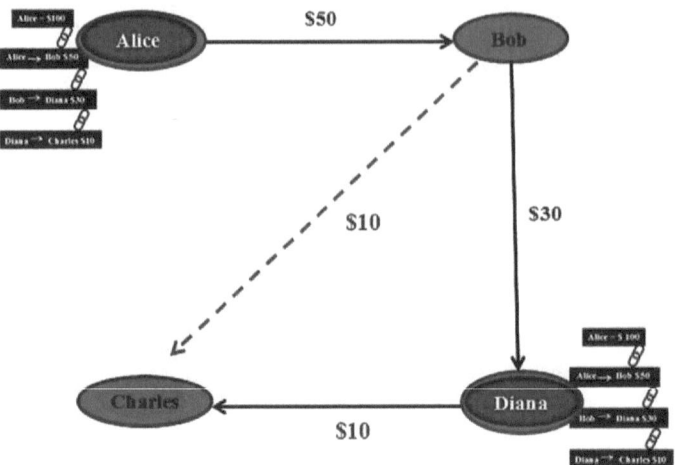

FIGURE 12.5 Synchronization of ledgers.

exploits the receptiveness of the record and make a huge number of companions and spreads fashioned exchanges, the whole framework would crumble right away.

In Figure 12.5, Alice and Diana are miners as they have a copy of the ledger.

To synchronize the records over the system, excavators will contend among themselves concerning who might be the first to take up the new exchange and approve it. The chief excavator that can approve and place it into the record will get a money-related prize. To win the opposition, to add the exchange to the record, a digger needs to complete two things:

i. It needs to approve the new exchange: This is simple as the record is open, and promptly figuring whether Bob has adequate assets so as to make the exchange is only a scientific count.

ii. The digger should locate an exceptional key that will empower it to take the past exchange and lock the new exchange with it. For this, excavators need to contribute their computational force and time since this quest for the key is arbitrary. The digger needs to more than once continue speculating new keys until it finds the one that coordinates a sort of an arbitrary riddle. If there should be an occurrence of bitcoin, they need to discover a hash esteem, an answer of a cryptographic capacity, which associates the new square with its antecedent. This is known as the proof-of-work [15].

The trouble of this work is chosen to confine the pace of age of new squares by the system to one at regular intervals. Because of the extremely low likelihood of effective age, this makes it unusual which excavator in the system will have the option to create the following square. In bitcoin, it depends on the SHA 256 hash calculation. The digger that succeeds will get the monetary prize generally as far as digital money is included. This is the best way to make substantial bitcoins. Bitcoins must be made if diggers explain the cryptographic riddle. As the trouble of the riddle builds, time and PC power contributed by the excavator increments. Subsequently, there is just a particular measure of cryptographic money token that can be made in a given measure of time. This is a piece of the accord no friend in the system can break. Litecoin diggers are as of now granted with 25 new Litecoins per hinder, a sum that gets split generally at regular intervals (every 840,000 squares) [11].

12.7.4 Synchronization of Ledgers

Only miners can confirm transactions. If a miner is able to solve the puzzle, it edits its own ledger first, by linking the current transaction to the previous transaction. It then publishes the solution to the entire network through broadcast. The solution indicates that the transaction is validated and the key enables everyone on the network to take it and add it to the copy of their ledger. All the other miners suspend the validation process and immediately take this broadcast message and add the transaction to its own ledger.

Figure 12.6 shows that Diana and Alice have a copy of the same ledger. The transaction is now complete and the miners move on to look for additional transactions to work on, confidently to get the reward next time.

12.7.5 Double-Spending Problem Solved by Blockchain

The inherent drawback with digital money is the double-spending problem, which is because of the centralized ledger arrangement. Double-spending indicates "spending the same money twice". If Alice has physical cash and once the cash is handed over to Bob in exchange of goods, the money goes to the Bob and cannot be used by Alice again. Digital money is easy to copy as it is a string of alphanumeric characters.

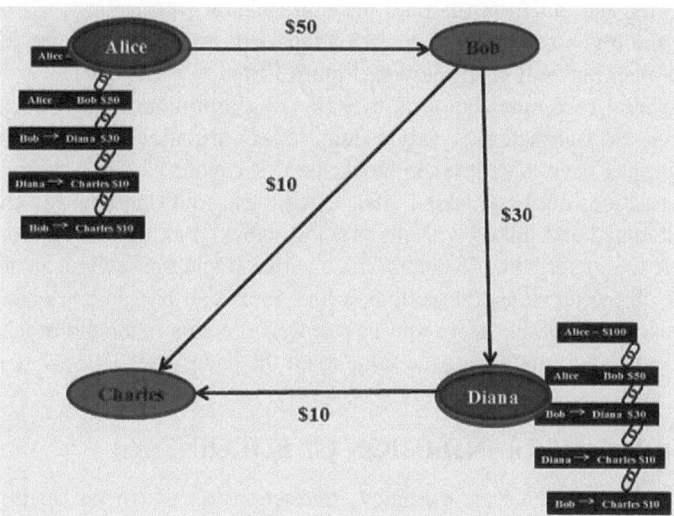

FIGURE 12.6 Synchronized ledger.

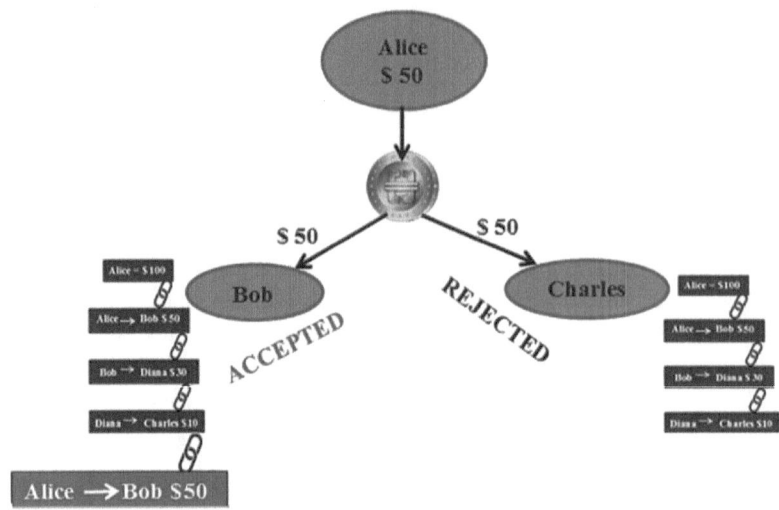

FIGURE 12.7   Solution to double-spending problem.

This means that it is possible to copy the transaction details and rebroadcast it such that the same money could be spent multiple times by a single owner.

Let us assume that Alice has $50, and she is attempting to transfer $50 to the wallets of both Bob and Charles simultaneously. Both these transactions would go into the pool of unconfirmed transactions. The miners now get into action and if the transaction to Bob is confirmed first, then the transaction is verified, confirmed, and linked to the previous block. However, the transaction to Charles would be recognized as invalid by the confirmation process and would be rejected. If both transactions are dragged out from the pool for confirmation concurrently, the transaction that gets the highest number of confirmations will be included in the blockchain, while the other one will get discarded (Figure 12.7).

The general recommendation is to wait for a minimum of 6 confirmations to ensure that the transaction is valid. Here, "6 confirmations" implies that after a transaction has been added to the blockchain, six more blocks containing several other transactions are to be added after it. All these confirmations and transactions are timestamped and linked with the previous block, making them irreversible and impossible to tamper with. To compromise with a single transaction, an attacker has to reverse all the previous transactions, which in case of bitcoin dates back to 2009. This is next to impossible as it requires a powerful computational power to alter the entire ledger and communicate the same to all the nodes.

12.8   REAL-TIME TRANSACTION OF BITCOINS

1. Merchants love crypto-currency transactions as it can be easily moved between two people for a very less transaction fees. The demand for cryptocurrencies have increased over the years because of the trust.

The transaction is similar to those of banks, where if someone gives you a check and you deposit it in your account, the money is reflected in your account. The equivalent to bank account in cryptocurrency transactions is a "wallet". In bitcoin, if someone gives you a bitcoin, it is stored in your wallet. The transaction is reflected in the ledger and it is there for everyone to see. There are different kinds of wallets: online, mobile, desktop, hardware, and paper wallets. With these types of wallets, money can dwell on one's own computer, on paper, or on online in wallets hosted by different companies.

2. There are two parts to the bitcoin address: the first being the public key, which is a series of 27–34 alphanumeric characters and appears as follows: M2810Gjs3001XsMrInr0204MatZg2a.

3. This can be thought of something similar to an e-mail address, which can be shared with anyone who wants to send cryptocurrencies to one's account. Security comes in the form of private key that is analogous to the password. Private key is required to access the account and view the chain of blocks from the public ledger. There are three methods to purchase bitcoins: the first method is direct purchase in exchange of fiat currencies. There are websites such as Coinbase, Robinhood, Binance, Cash App, GDAX, Coin Exchange, Coinmama, Bitstamp, to name but a few. These are locations similar to stock market, where people buy and sell bitcoins. The second way to acquire bitcoins is through exchange of goods or services for bitcoins. Irresistible discounts are offered if the purchase is through bitcoins. This is through the understanding that due to its limited edition, the value of bitcoin is bound to have a steep increase compared to fiat currencies. For example, a dealer would be ready to trade off a mobile phone worth $80 for bitcoins worth $50 as he is bound to strike a better deal in using bitcoins. People have started buying homes, Lamborghinis, and Ferraris using bitcoin. The last method is earning bitcoins for free, through mining. Bitcoins are released every 10 minutes, and there are lots of miners who are on the lookout for a new block to be created, solve the puzzle, and get rewarded with bitcoins. This depends mainly on the CPU power and the Internet speed of the miner. Though it sounds easy, it is getting more difficult with the increasing number of miners. To succeed, one has to invest in expensive hardware and electricity consumption.

12.9   CRYPTOCURRENCY MARKET

Cryptocurrency market is similar to share market, where the statistics of all the cryptocoins are presented. The overall *cryptocurrency market* is projected to reach US$1.40 billion by 2024, at a CAGR of 6.18% during the forecast period [16].

- Create Coinbase account
- Convert your money to bitcoin
- Create Bittrex Account
- Trade your bitcoin for other cryptocurrencies

12.10 PITFALLS IN CRYPTOCURRENCY

Cryptocurrencies have a big risk in the form of government or political intervention. But an even bigger risk is bugs in the code or problems with the protocol that leads to a complete lack of confidence. Though it is structured to overcome the weaknesses of trust-based models, the entire network is built on mutual trust among the peers. If trust is lost, then the entire network would crash to almost zero.

12.11 RISK AND FUTURE OF CRYPTOCURRENCY

Bitcoin was just an experiment, which turned out to be successful. As of now, cryptocurrencies hold a high value, but it also has been quoted as "the big bubble". This happens when people flock up to a new investment vehicle that drives prices up like a "tulip mania". The biggest question is: "Will bitcoin become widely accepted as a form of payment and a store of value?". In other words, will the mass majority of people in the world use bitcoin as a means of exchange over other currencies. The first bubble was witnessed in June 2011, when the price of bitcoin soared from 50 cents to 32 dollars in about 60 days and then over the course of a few months back to $2. Another big bubble was observed between April 2013 and November 2013, when the rates went up from $35 to $266 and crashed back to $50. All these are small blips on a map as the exchange rate of bitcoin to date is $8,725.77 (according to Coinbase accessed on 21 January 2020). The future market is unpredictable, but there's an old cliché of Warren Buffett that says: "when people are scared, be greedy and when people are greedy be scared". It applies to every market and especially cryptocurrencies. Investing in cryptocurrencies using money that we cannot afford to lose brings in significant risks.

12.12 CASE STUDIES

12.12.1 CURRENCY CRISIS

ZWD is the cash condensing for the Zimbabwe dollar, which was the official money for the Republic of Zimbabwe from 1980 to 2009. In 2007–2008, ZWD experienced one of the most exceedingly awful scenes of hyperinflation at any point recorded. By printing much cash in the range of 2 years, the Zimbabwe government basically downgraded their money to nothing. This prompted a cash emergency, which realized a sharp decrease in the estimation of the money. Such decrease in esteem makes individuals pull back the entirety of their cash and convert it into outside money. This is alluded to as capital flight. Subsequently, the trade rates deteriorate, bringing about a sudden spike in demand for the cash. Two other mainstream money emergencies are Latin American Crisis of 1994 and Asian Crisis of 1997. Following quite a while of hyperinflation, the legislature of Zimbabwe reported the demonetization of the ZWD in 2009, which got last in 2015. Demonetization is the procedure of formally expelling the lawful status of a cash unit. National banks are the principal line of protection in keeping up the steadiness of money. Bitcoin is inalienably intended to be deflationary to control expansion. It handles this issue by foreordaining and controlling the measure of bitcoins that will be discharged throughout the years.

12.12.2 Scenario Subsequent to Mining All the Bitcoins

Expecting that there are no progressions to the current bitcoin convention, the bitcoin top of 21 million BTC will come by 2140. As of December 2019, an aggregate of 18,105,438 BTC has been mined (https://www.blockchain.com/en/diagrams/absolute bitcoins). This number changes about like clockwork when new squares are mined. At present, each new square includes 12.5 bitcoins into dissemination. At present, diggers acquire a large portion of their salary using the square prize. At the point when each of the 21 million bitcoins is mined, there won't be a square compensation to pay to diggers. At the point when a bitcoin client sends a BTC exchange, a little charge is appended. These charges go to diggers, and this is the thing that will be utilized to pay excavators rather than the square prize [15]. During this time, swelling will go down to 0%.

12.12.3 OneCoin Trick Is as Yet Alive

OneCoin was advanced by Ignatov, a Bulgarian national, as a cryptocoin with a private blockchain. OneCoins are created through mining and could be utilized to make installments for items advanced by OneCoin. The coin additionally accompanied a digital currency wallet and was evaluated that a sum of 120 billion coins would be accessible on the OneCoin system. Coordinators sold instructive materials, including limited time and limited bundles to organize members. In evident staggered showcasing style, members were additionally allured with referral compensations to encourage more clients to join.

A few reports recommend OneCoin was a Ponzi plot and is confronting legitimate activity in a few countries around the world. A few able specialists of European Union nations caution of the dangers that may emerge from OneCoin. For example, UK's Financial Conduct Authority (FCA) recommends that OneCoin is being researched by the police, the Financial Services, and Markets Authority (FSMA cautions that neither OneCoin nor the people advancing OneCoin have been perceived or approved by the FSMA). The Hungarian national bank (Magyar Nemzeti Bank, MNB) recommends that this may be a fraudulent business model and cautions of the potential dangers that might be related to it. Konstantin Ignatov was captured in March 2019, on charges of a wire extortion trick originating from his job as the pioneer of a worldwide fraudulent business model that included the showcasing of "OneCoin". A prosecution charging Ignatov's sister, the originator and unique pioneer of OneCoin— with wire misrepresentation, protections extortion, and tax avoidance offences, has been enrolled. Because of deceptions made about OneCoin, unfortunate casualties have lost billions of dollars worldwide in the fake digital currency. Despite the fact that OneCoin is broadly known to be a trick, the plan said it denies any bad behavior and isn't answerable for its "self-employed entities" who sell on its items.

12.13 CONCLUSION

Mixed opinion prevails about the future cryptocurrency market. However, with the world economies dwindling, this market provides an alternative venue for transaction and investment. The strong point is that most of the cryptocurrencies are based on blockchain technology, which creates a more trusted environment. Though blockchain wins over all the prevailing technologies, not all businesses can adopt blockchain. The impact and utilization is an open debate, but with no alternative as of today, this technology scores over all the existing ones. The prospects of fading as a bubble or evolving as a life-changing foundation of most businesses will be evident once the system matures and is accepted by the masses.

GLOSSARY

Standard medium of exchange: Money is to be fungible or interchangeable, in the sense that the currency can be exchanged for another equivalent value. One Canadian Dollar is equal to Rupees 54.32.

Store of Value: Money should be durable be able to stand the test of time. For example, if I have a $ 100 bill in my desk, it will not evaporate or disappear into thin air with time.

Unit of Account: Money should be divisible or broken down into smaller parts. Eg. Dollar-Penny – cents, Rupee – paise, etc.

Wallet: A wallet is a software package that stores private and public keys and interacts with various blockchain to enable users to send and receive digital currency and monitor their balance.

REFERENCES

1. "Top 100 Cryptocurrencies by Market Capitalization", retrieved from https://coinmarketcap.com/ on July 2020.
2. https://hackernoon.com/why-everyone-missed-the-most-mind-blowing-feature-of-cryptocurrency-860c3f25f1fb
3. Duffy, John (2007). *Cliffs Quick Review Economics*. Houghton Mifflin Harcourt, Wiley Publication Inc., p. 63.
4. ""Currency". *The Free Dictionary*", retrieved from https://www.thefreedictionary.com/currencies on July 2020.
5. Bernstein, Peter (2008) [1965]. "4–5". *A Primer on Money, Banking and Gold* (3rd ed.). Hoboken, NJ: Wiley. ISBN 978-0-470-28758-3. OCLC 233484849
6. https://thenextweb.com/hardfork/2019/02/19/the-differences-between-crypto currencies-virtual-and-digital-currencies
7. https://appiqotechnologies.wordpress.com/2018/02/12/why-should-you-invest-in-cryptocurrency
8. "Buy and sell cryptocurrency", retrieved from https://www.coinbase.com, on Jan. 22, 2020
9. "Cryptocurrency", retrieved from https://www.investopedia.com/terms/c/crypto currency.asp on July 25, 2020.
10. "Ethereum", retrieved from https://ethereum.org on July 25, 2020

11. "Lite coin: the cryptocurrency for payments", retrieved from https://litecoin.org on July 25, 2020

12. Markovich, Sarit (5 December 2017). "Commentary: The Overlooked Actor That Could Crash Bitcoin". *Fortune*. Archived from the original on 5 December 2017.

13. Popper, Nathaniel (13 June 2018). "Bitcoin's Price Was Artificially Inflated Last Year, Researchers Say". *New York Times*. Retrieved 13 June 2018.

14. Vigna, Paul; Russolillo, Steven (12 August 2018). "The Mystery Behind Tether, the Crypto World's Digital Dollar". *Wall Street Journal*. Retrieved 17 August 2018.

15. "Bitcoin", retrieved from https://en.bitcoin.it/wiki on July 25, 2020

16. https://www.marketsandmarkets.com retrieved on July 25, 2020

13 Emerging Trends and Research Issues on Blockchain Technology for 5G-Enabled Industrial IoT

K. Suresh Kumar and V. Kishore Kumar
IFET College of Engineering

T. Ilamparithi
Manonmaniam Sundaranar University College-Puliyangudi

S. R. Boselin Prabhu
Surya Engineering College

R. Dinesh Kumar
Siddhartha Institute of Tech and Science

CONTENTS

13.1 INTRODUCTION

Bitcoin and Ethereum are perhaps the two most generally perceived usage of the blockchain. The distinction between Ethereum and bitcoin is that bitcoin is just money and Ethereum is a record innovation that organizations are utilizing to fabricate new projects. Both bitcoin and Ethereum work on what is classified as "blockchain" innovation. Ethereum is unquestionably increasingly secure. It is a cryptographically connected and persistently developing rundown of permanent information records. Within the blockchain, a (open) record is utilized for recording the information, just as the data of each exchange. Data about each finished exchange are put away in a dispersed record and shared to all the participating hubs of the blockchain arrangement. As talked about in writing, blockchain can effectively record exchanges at least between two gatherings included on a conveyed distributed (P2P) arrangement, with the put-away information co-possessed by all individuals from the system and for all time modifiable [1]. In the end, blockchain gives a permanent, trusted, and ensured stage for different elements (the two people and associations) to trade information/ resources, team up, and perform exchanges.

The Internet of Things (IoT) is empowered by means of recent improvements in the radio frequency ID, communication technologies, Internet protocols, and smart sensors. One of the fundamental evidences is to have the intelligent sensors collaborate with and without human participation for delivering a recent level of applications [59]. The blockchains are significant structure that obstructs later on an interconnected and progressively mechanized world. They are getting a charge out of fast development all alone [60]. The crossing point of these two territories makes intriguing use cases and attendant issues. A convincing explanation impelling the collaboration among these dualistic regions is indicated by the absence of protected strategies to mechanize trust and trade off ongoing information concerning IoT gadgets, and blockchain gives a practical arrangement [61]. A significant test popular in the region of blockchain is that it is a moderate innovation and has not seen far-reaching reception across various ventures. Therefore, there is an absence of scholastic articles that might light up the subjects encompassing blockchain appropriation. The rising application territories involve cooperation between IoT gadgets and block-chain as well as some recent innovations and research trends in this technology. The accessibility of IoT-enabled sensors might be utilized to build up attribution and empower the accomplishment of shrewd agreements. With regard to vitality of net-works, early examinations are in process of utilizing blockchain for keen metering and directing dispersed vitality exchanges. We picked these three significant territories because they see expanding examination, venture, and development. Blockchain utilizes circulated and open records to empower mysterious exchanges that can be trusted [62]. Significant issues related to blockchain discourses are the intermediation wherein purchasers are coordinated with dealers through an outsider both confided in, generally a bank or a representative in money-related exchanges. The greater part of the current exchanges includes a brought together model [63]. Blockchain offers a component to expect the job of this delegate [64] and moves us away from an incorporated to a deunified model. Each gathering can confirm the exchanges of different gatherings through the blockchain. The massive size of the IoT includes billions of gadgets that might need to direct exchanges between one another. A significant test coordinates the billions of gadgets. An incorporated model isn't pertinent here; hence, blockchain offers a decentralized arrangement [65]. Blockchains additionally offer a few attractive capacities, including value-based legitimacy, value-based perseverance, and value-based security [66]. They address the issue of information power, where people are given authority over their information and can impart it to just the gatherings they trust. We will analyze these attractive highlights of blockchains related to the three center application territories we decided to address right now. Since there are numerous fantastic audit articles, texts, and magazines clarifying the fundamentals of blockchain, we give a concise diagram to keep the phrasing right now contained. The peruser has alluded to the accompanying hotspots for portrayals of the blockchain and prior reviews of blockchain for IoT [67,68].

13.2 BLOCKCHAIN APPLICATIONS: INDUSTRY 4.0 AND INDUSTRIAL IoT (IIoT)

The use of blockchain technology in IoT (i.e., IoT-enabled industries such as IIoT and Industry 4.0) is predicted to develop, and it has benefitted to a wide variety of sectors. Many techniques like resilient techniques are used for record-keeping, and they offer great game changers by this technology, just like Tamper-proof (the same technology). Many cases have been noted and studied. It is known that there have been certain investments in the blockchain. Rather than considering the issues related to the scalability, the rational method of data sharing, which is a centralized one, is considered to give more safety than the decentralized one [1].

13.2.1 Industry 4.0

Industry 4.0 was introduced in Germany in 2011, and since then it has made a massive improvement; it has been upgraded to Industry 4.0 [2]. In the initial stage, it emerged by getting distributed over by some technical aspects, and the networks with dynamic capabilities got highly automated with some essential technologies driving them: Internet, Big Data analytics, edge/cloud computing, etc.. The industrial systems with automated technology are internally connected systems (cyber-physical systems) supporting industry 4.0 by permitting the industrial arrangement. The construction process is allowed to transform into a dynamic system and is highly automated [3]. The individuals in a highly integrated network should be communicated and it behaves like an intelligent device in order to work autonomously with each other and to reach a common goal [4]. The data and correspondence advances (ICT) remain foreseen on the way to show strategic jobs in reasonable development to help worldwide financial, social, and natural manageability [5].

Industry 4.0 consumes three ideal focal models. The principal worldview exists as a shrewd item that proceeds with a control of the assets and organizes the assembling procedure to its extreme end. The subsequent worldview is that the shrewd mechanism is a digital to physical framework, where customary fabricating forms change into circulated, versatile, adaptable, and self-sorting out creation lines. The third worldview is the increased administrator, which includes adaptability also, the ability of a human administrator in the mechanical framework. Innovations will, at that point, be utilized to help the human administrator, recognizing its centrality when confronted with an assortment of occupations like determination, observing, and check off the creation forms. It targets enlarging laborers' capacity and giving a friendly workplace, reshaping their job underway cycles over and done with human–machine interfaces empowering cooperation amongst the whole producing biological system. The Internet is the furthermost undeveloped innovation of Industry 4.0 as a lion's share of the other innovation drivers of Industry 4.0 stay reliant on it. For instance, the continuous distribution of data concerning different substances through a computerized correspondence system or utilizing figuring force and information capacity ability of distribution center scale PCs remotely is empowered through the Internet. Cleverness in a framework is accomplished through the coordination of articles, items, and administrators and giving setting mindfulness by means of the Internet [6].

13.2.2   IIoT

IIoT empowers the combination of remote sensor systems, correspondence conventions, and web framework with the procedures empowering clever mechanical tasks for checking, investigation, and the executives [7]. This between the systems of anything in the creation framework setting helps in the robotization of mechanical creation furthermore, improves insight, productivity, and well-being. The IIoT architecture has three layers: physical, communication, and application. All layers might have certain functions as like IoT. The major application in the IIoT is that it is used in 5G network-based automation. IIoT can fundamentally change the business as we probably are aware of these days, yet symptoms can't be belittled. Security, trust, and altering will become serious issues in the fourth and fifth modern insurgencies; besides, it isn't unimportant to tackle them in a setting where low-controlled and resource-compelled gadgets are omnipresent. This leads to confide in issues [8] that the blockchain frameworks are attempting to manage. In Reference [10], a confirmation component in light of blockchain-empowered mist hubs has been executed utilizing Ethereum shrewd agreements for the requirement of IoT gadgets security.

13.2.3   Blockchain: 5G-Enabled IoT

IoT associates billions of articles together for fast information move, particularly in the 5G-empowered modern condition during data assortment and handling [11]. This brought-together design may have a solitary purpose of disappointment alongside the computational overhead. Along these lines, there is a requirement for a creative dispersed access control section for communicating with other devices in different modern areas such as IoT-empowered mechanical robotization. In such a domain, security and protection safeguarding are significant worries as a large portion of the arrangements depends on the concentrated engineering. The approach of 5G technology and distributed computing resolve improve the limit, usefulness, and adaptability of a system administrator to propose the new scope of administrations [9]. Information examination on the gigantic measure of information gathered by the researchers from the available massive count of operable sensors with a variety of applications in the IIoT use cases can be overseen in the financially savvy way by 5G and computer-based intelligence-based information investigation, because of its high information speed and low dormancy guarantees. 5G arranges requirements to utilize blockchain innovation to convey secure correspondence. IoT indicates the system of different particular gadgets (both electrical and electronics), which can fit to collaborate with one another utilizing the open channel, for example, the Internet. The association is thus made utilizing remote innovation, for example, sensor systems, radio recurrence recognizable proof (RFID), close to handling correspondence (NFC), M2M, and ZigBee [12,13]. At that point, IoT has changed the domain of omnipresent registering with various mechanical applications that worked with different kinds of sensors. In any case, there are definitely constraints with the use of IoT, which should be settled down so as to develop it into an increasingly proficient framework [14,15]. Increments in the quantity of IoT-empowered gadgets are responsible for a requirement for an innovation that can bolster this tremendous measure of information transmissions effectively at an amazingly high data transfer capacity. In addition, the gadgets themselves must have the option to deal with these changes in arrangement, for example, huge transfer speed limit and improved information rate with low latencies [16]. The appearance of quicker remote advancements, particularly the fifth era remote frameworks (5G), is a boosting driver for the 5G-empowered IoT applications. It additionally assists in taking care of a massive number of IoT-empowered gadgets [17].

13.2.4 Blockchain: 5G-Enabled Smart Industrial Automation

The industrial applications in the blockchain-based 5G IoT are given in the form of taxonomy that covers some of the areas like healthcare, industry, agriculture, smart homes and cities, autonomous vehicles along with the supply chain management schemes, etc. In all these areas, the 5G is combined with a blockchain for improving security. It is also useful for boosting the bandwidth, also limiting the complete operational experience as well as the capital expenditure.

13.2.5 Blockchain-Based Smart Homes Using 5G

The smart home is an automatic inclusion of highly innovative and advanced living conditions that are developed to expand the individual fulfillment of the inhabitants [18]. It provides much security, space to live, and space to the owners, thus allowing control over the settings as directed by the deposition by the assistance of an innovative unit application. In IoT, the smart home–related frameworks will enable the usage of the gadgets that are used to systemize some of the various activities that depend on radical measurements to provide stable and continuous managerial support as per the client's interest. Some research initiations are discussed with reference to the idea about the vitality of the insightful home. The generic framework starts from a smart home comprising assets that are accompanied by WiFi connection along with IoT-enabled devices, which are used for some home-related works and various applications, as mentioned for remote access [19]. The fundamental directions given by the smart homes might accompany with lights, locking systems, accessories regulators, video exploration, and intense stopping [20]. Together, these will give the needed experience ideally to the occupants of the home. The various directions should ceaselessly trademark the pertinent information and data that can coincide most effectively. The most intended entryway framework is a necessary part of the smart home. The essential aim is to recognize the unauthorized and unauthenticated clients entering the house. The details of the persons inside the home are stored in the cloud server to approach home-automated devices. Also, it provides authentication-based security so that if the data handled by the system undergo any failure or if any outsider hacks the system, then the system gets locked. This is handled with the introduction of a blockchain-based lock system having some additional features that enable the system to handle the insiders/outsiders effectively. The information on the whole transactions that have open or lock command will be used to store as network blocks in the blockchain. However, the blockchain network makes any intruder obtain unauthorized access to the system, and that supports in changing the existing executed businesses. These kinds of problems can be easily handled using 5G wireless technologies.

13.2.6 BLOCKCHAIN-BASED SMART CITIES USING 5G

The expanding pattern of individuals to relocate to urban regions combined with the related procedure of urbanization causes numerous mind-boggling difficulties in regard to the urban communities' general foundation and their capacity to give residents the essential amenities like water, vitality, transportation, and medicinal services. This uncommon urban development is because of factors such as climate change, increment in populace, and shortage of assets [21]. A review of IoT-based shrewd urban communities features different applications, advantages, and weaknesses [22]. These were discussed and analyzed to incorporate the application in the field of blockchain technology; this was wisely incorporated into some savvy urban areas, for example, the "Padova Smart City" [23]. A fundamental part of shrewd city is insightful stopping framework, which helps in the improvement of traffic, and the board frameworks to decrease the expense caused by employing significant staff; for instance, a calculation that expanded the effectiveness of the cloud put together brilliant stopping frameworks based on IoT innovation [24].

13.2.7 BLOCKCHAIN-BASED HEALTHCARE USING 5G

Social insurance is one of the most essential angles for the general advancement of any country. It tends to be considered as a sign of a public's general prosperity. With the increase in population and ailments, the pressure on present-day medicinal services frameworks has increased. 5G-empowered IoT considered as a potential answer to mitigate the pressure on social insurance framework [25–27]. For instance, a recognized vital segment with lot of health-related documents might get started to accomplish IoT-based social insurance framework for remote checking of strength of fundamentally sick patient [28]. Individual well-being record (PHR) [29] is identified with the advanced record of an individual patient. EHR empowers secure, ongoing sharing of clinical treatment narratives of patients to specifically approved clinical workforce [30]. A decentralized record of the executive's framework named as MedRec to deal with EHRs utilizes blockchain innovation. It handles private data and oversees essential contemplations, for example, verification, secrecy, responsibility, and information sharing. It empowers clinical partners, for example, general well-being specialists, analysts, and specialists to partake in the blockchain to organize an "excavators" to give specific motivators [31].

13.2.8 BLOCKCHAIN-BASED INDUSTRY 4.0 USING 5G

Right now, complete computerization of modern and business forms has become a reality. Huge improvements in innovation and their presentation into the industry has brought about the rise of another way to deal with creation, known as Industry 4.0, such as IoT, blockchain, and cyber-physical systems [32]. In Industry 4.0, IoT depends upon the promising transformational answers for existing modern frameworks. In this way, it is viewed as a key empowering agent for the upcoming age of cutting-edge modern robotization [33]. Because of the profoundly serious market, organizations intend to pick up business focal points at any expense. These power businesses process executive's frameworks in Industry 4.0 to digitize and computerize business procedures to build their benefits. Notwithstanding, by adding independent operators to these business forms, the exchange expenses and dangers related to them will also increment. A potential answer to handle these dangers is that every specialist must discuss straightforwardly with one another. It tackled the issue of exchange costs for independent specialists. In any case, there emerges an issue of trust between these participating operators. So, to handle all previously mentioned issues, the decentralized frameworks (blockchain innovation) must be utilized and the correspondence between the independent specialists in a multi-operator framework must be ensured [34,35]. The Autonomous Intelligent Robot Agen (venture AIRA) [36] represents a standard of financial association between human-specialist and operator. On the comparable line, an investigation is conducted that creates the chance to execute the mechanization in BPM frameworks utilizing blockchain innovation. Utilizing blockchain in business process guarantees the activity of administration with trust and security among the included gatherings [32].

13.2.9 BLOCKCHAIN-BASED SUPPLY CHAIN MANAGEMENT USING 5G

The supply chain is the system of people, associations, assets, and exercises that are engaged with the existing pattern of an item. It begins from article creation to its deal, from the supply of crude materials from provider to producer, to the end client. The standard stream in a production network starts with the provider, trailed by the producer, distributor, retailer, and, lastly, to the purchaser [37]. Given the worth of supply chains, it additionally faces some difficulties [38]—the effect of two advancements on inventory network: blockchain and 5G-empowered IoT. 5G-empowered IoT expands the transfer speed limit of the transmission of products-related information. The blockchain gives an unchanging, conveyed record that empowers secure capacity of data. In addition, it may very well be utilized as an instrument to forestall the event of malignant IoT gadgets into the system [17]. Other than the prudent effect of blockchain innovation on organizations, it can conceivably assist with moderating legitimate charges emerging from questions. The primary part of blockchain innovation is a brilliant agreement that can empower programmed installment of products upon their receipt; in this way, the requirement for an outsider affirmation is removed. Another significant viewpoint is to dispose of questions concerning whether a merchant is qualified for a volume-motivating force discount. This can be dealt with by utilizing savvy contracts combined with 5G to follow a shipment.

13.2.10 BLOCKCHAIN-BASED AGRICULTURE USING 5G

The discernibility of agriculture nourishment store network of the board is essential to guarantee the sanitation. It likewise builds consumer loyalty and distributed profitability. The incorporated information stockpiling makes it increasingly hard to guarantee the quality, rate, and cause of the items. So, we need a decentralized framework where straightforwardness is accessible, which makes individuals from the makers to buyer's fulfillment. The innovation of blockchain is an advanced one that permits us to get discernibility and straightforwardness in the production network. Utilizing this innovation improves the system between various partners and ranchers. The properties of blockchain basically give expanded limit, better security, unchanging nature, printing, quicker settlement, and full recognizability of put-away exchange records [69].

Smart agriculture utilizes present-day innovations for improving the value and amount of agricultural products derived as outputs. The information gathered on temperature (in Celsius), light (solar or external sources), quality of the soil (including the nutrients content), and mugginess (air humidity) can be placed in a pivotal control framework and dissected utilizing certain artificial intelligence calculations [39]. The amalgamation of different advancements in brilliant farming means to make the rural inventory network financially savvy with no trade-off in the item quality. Disseminated Ledger Advances (DLTs) are considered as the best potential device to expand the proficiency and straightforwardness in this blockchain-based agriculture [40]. The most basic perception that DLTs give is upgraded discernibility. Thus, it could follow businesses that happen all over the supply chain in real time.

13.2.11 BLOCKCHAIN-BASED AUTONOMOUS VEHICLES USING 5G

The advancement in technological developments makes a rapid variation in recognizing, investigation and identification as well as computation and communication. The development of the Intelligent Transportation System (ITS) is challenging to process thoroughly. ITS consists of safer, smarter, and most convenient facilities that are used for transport and offered services. A severe security risk exists because of centralization that leads to centralized authorities down the system temporarily to convinced peripheral attacks from malicious websites. Furthermore, the absence of sufficient conviction among the transportation systems-related specialists should be handled. To defeat the previously mentioned difficulties, a blockchain-based rule is made to ensure trusted and decentralized self-sufficient environment for ITS. The soaking stone for Parallel Transportation Management Systems (PTMS) structure will enhance these present reality transportation frameworks utilizing equal connections with their partners [41].

13.2.11.1 Blockchain-Based Unmanned Aerial Vehicles (UAVs) in IIoT Using 5G

The UAV is an airborne framework or a remote-controlled device that is used for blockchain-based designs [42]. The pictures acquired from UAVs can offer help in different modern applications, for example, urban demonstrating, observation, enormous scope mapping, conveyance, correspondences and media, search-and-salvage activities, and agribusiness [43]. UAVs are worked by a few military powers and furthermore by certain regular citizen associations. The greater part of the business employing UAVs relies upon the Wi-Fi network to be remotely open. Be what it may, the Wi-Fi network isn't adequate for the visual view (LOS) correspondence. Instead, omnipresent portable systems, for example, 5G, can be utilized to work with the LOS correspondence. Additionally, they offer a wide-region, rapid, and secure remote availability that might improve the controlling and well-being of the gadget [44]. With the ascent of self-ruling UAVs, dangers of interruption or capture attempt have increased. An enormous number of assaults target such UAV systems; for instance, to stick the correspondence arrange, infuse false information, and upsetting the system tasks [45]. The actualized digital security framework depends on Intrusion Detection Systems (IDS) to defend UAVs from outside digital assaults [46]. At present, each UAV can screen the conduct of its patrons. If an IDS operator is suspected to be malevolent, it is banned from working as an observing hub.

13.2.12 Blockchain-Based Multimedia and Digital Right Management Using 5G

Media conveyance is an important type of computerized circulation of sight and sound substance, for example, sound, picture, and video [49]. Contrasted with the past stages to convey media, for example, minimal plates, conveyance medium, for example, peer-to-peer network or cloud administration, has developed the present standard for interactive media conveyance. The benefits of the presently accessible substance for video conveyance medium incorporate high accessibility, cost viability, and better quality. The customary CDNs are attributable to their lower lodging cost, since owning of the framework is not required. Besides, these frameworks have certain characteristic issues that are hard to determine, given the unified design of current Digital Rights Management (DRM) frameworks [48]. A blockchain-based substance dissemination framework helps the media makers request a proficient route for computerized rights to the board for the effective transmission and the media holders will have the ability within themselves to work with this framework [47]. This structure supplements the current DRM frameworks, where the latter requires a verification component, which is given by the previous, utilizing blockchain. In any case, being an underlying model, the main disadvantage is that it had no motivating profit model, so the excavators cannot get any prize for the effective mining of a block.

13.3 BLOCKCHAIN: OPEN ISSUES AND CHALLENGES

For the scholarly community and industry, the IoT and blockchain have become impressive research intrigue. The PoW idea based on bitcoin-style blockchain has a few attributes that make it ineffectively appropriate for some IoT situations. In light of the broad writing audit, challenges talked about with blockchain-based 5G-enabled IoT applications, and issues outlined, we have distinguished significant research identified with blockchain-based 5G-empowered IoT in the form of IIoT. Consequently, these issues and open challenges are to be discussed further, along with existing frameworks to identify future trends in the blockchain. The technology that keeps on changing its direction based on persistent revolution is "Blockchain Technology". It has trusted information stored in the chain of blocks. An overview of information on emerging trends in the blockchain-based technology for healthcare applications is given and the prospective of this technology in the medical care industry is shown.

The information will be empowered and streamlined by the sharing process of the health data of a patient. Blockchain technology will hold the data of patients with controlled rightful owner data and also give rights to the patients for deciding the person who has the authority and responsibility to access their data intended for some purpose [50]. A large number of records of a patient in the digital form need to be generated because of the upcoming digitalization process. Because of this digitalization process, a massive quantity of data protection is needed that may lead to the existence of the blockchain technology. This blockchain technology delivers safety and data privacy, along with the eradication of integrity issues. Various research and applications on the blockchain technology provide a new technology that can be used for the end-to-end communication of digital data that can be transmitted privately or publicly to all the users. This technique also provides the storage of data in any needed format. This technology provides reliability and transparency and avoids intermediaries or third party administrators. This makes uses of mechanisms to verify the transaction in a trustless and unreliable environment [51]. New technology has been proposed with high security, excellent privacy, along with very light architecture, which is well-suited for IoT-based blockchain technology for avoiding the disadvantages of blockchain for security and privacy. A brief description of the method is given, which can be used for the smart home application using the IoT applications. This hierarchical architecture, which is proposed in this chapter, contains smart homes, overlaying network, and storage in cloud servers for transmission of data along with blockchain for providing security.

The designs specified in this chapter make use of the trust method in different topologies, which depend on different types of the network used in this blockchain. The effectiveness of the blockchain architecture in security and privacy for the mentioned IoT application is measured by using qualitative evaluation [52]. In the financial market field, for the purpose of transmitting data, the blockchain technology is initially designed. And because of its unique characteristics, the same technology is used for healthcare sectors. The main use of this blockchain technology in the healthcare industry includes security, easy access to health-related data and its data storage, efficient medical supply chain, and its different payment mechanism. It also acts as an essential method for the community behind the healthcare concept that has a huge amount of potential impact on the upcoming medical care applications. Here the importance of blockchain technology is suggested whose revolutionary improvement is going to create a new healthcare infrastructure that can be used globally. This blockchain technology deals with the issues that are related to the currently existing systems. It includes the absence of interoperability of electronic well-being record (EHR) programming, the wasteful and shaky exchange of ensured well-being data (PHI), an incapable installment framework for esteem consideration, and the advancing requirement for quiet, focused consideration.

The creation and execution of the blockchain technology in the healthcare field can overcome the problems in the current methods. Based on the utilization and effective usage by the clinicians, the modern approach can be made successful and will give a platform for the implementation of new ideas [53]. A new form of technology that is known as the revolutionary technology in the field of industry and commerce is the blockchain technology. The information about the currently booming multibillion-dollar global consumer electronics industry and its revolutionary impact is discussed. The efficiency of this blockchain technology is clearly mentioned in terms of transparency, safety, and honesty. The supply chain management areas of the consumer electronics industry will have a major impact of blockchain. Many initiatives and start-ups are in progress for the use of this consumer electronics industry. With new dimensions of this blockchain technology, the consumer electronics industry is benefited a lot. From this chapter, it is clear that blockchain technology has a great potential for providing transparency and traceability of the chain supplied along the time-constraint, i.e., the time required for applying the blockchain technology to all the areas of the consumer electronics industry. Due to the nascent stage of the consumer electronics industry, a lack of awareness about the blockchain exists among the employees in and around the business area. Applying the blockchain technology in many areas of the consumer electronics industry needs the fullest cooperation from the stakeholders and top-order representatives, to meet the real challenge in technological point [54]. Blockchain is used for storing the transactions of digital communication, initiates changes in innovation in the field of smart cities, e-Health, or e-Government because of its decentralized, immutable, and verifiable nature. But blockchain is subject to factors like scalability, security, and privacy issues such as crypto-key management and likability. To be a solution to this issue, blockchain with the crypto-privacy control technique empowers the users to take control over their personal information during each digital transaction of any kind. The current state on privacy preserving, the solution, and the different mechanisms in the blockchain technology are discussed in this chapter. The survey includes privacy-preserving research, and its solution by means of blockchain technology is being analyzed for various fields like e-health and e-government. Based on the related issues and various research challenges that are related to the privacy of blockchain, many different methods were identified and are deployed for the many scenarios like smart cities, e-health, and e-government [55].

The main and the most important characteristic of blockchain is its distributed behavior because of which centralized control is assigned and by getting approval, each transition is being carried out and the transaction is monitored regularly. This process guarantees that the transmitted data will not be easily traced back. There are many applications in which this blockchain is being used in which the IoT is a field that provides the fast-growing platform. This rapid change in the field of IoT based on the increasing demand in protecting all the connections used in the infrastructure from the information attacks and physical tampering is essential, for which the blockchain technology is mainly used that ensures security. An effective, new, and more scalable way of ensuring the data transmission and its entire process involved is the blockchain-based decentralized cryptographic process. This technology is going to be a method that can act as a backbone for IoT [56]. There are certain forms of study that identify the characteristics and properties of blockchain technology, through which open science is benefited. The requirement of the infrastructure of the open science has been estimated first, and then it is compared with the characteristics of the blockchain technology to estimate the needed characteristics related to the infrastructure. Blockchain technology will act as a starting point for the researchers to perform long-term research in the field of open science. This work identifies and compares the different characteristics of the open scientific infrastructure along with the characteristics of the blockchain technology and tells whether the blockchain technology is suitable. The result of the research suggests that the reliable and appropriate infrastructure of open science is well-suited to the blockchain technique [57].

In the recent analysis, the cryptocurrencies and the blockchain technology along with research challenges and opportunities are classified. This will give a clear idea for the researchers to work further in the field of blockchain technology. A guideline about how to build distributed ledgers is given, and it enables the software developers to know about the structure of the blockchain clearly for the development of digital currencies without the need or permission of central financial institutions. A distributed method that is shown in this chapter allows the blockchain technology to enforce other principles on the digital assets [58]. The basic types and design of blockchain for the IoT application as well as several challenges and issues that are related to the blockchain and IoT application are discussed. The future integration of blockchain with the IoT will provide an improved solution in the arena of security and privacy, which is tested by using a literature-based method, especially for the IoT applications. Some additional information to the researchers in the field is integrated with blockchain and the IoT with security and privacy [70].

The critical security issues have been studied in this research article. The researchers state the solutions for different security issues, and the theoretical solutions of the same methods are compared and discussed. This discussion will give direction to the research scholars who are interested in this area. The solution discussion will provide a new solution by combining the pegged side chains and the two-factor authentication. It is useful in improving the privacy of the blockchain and to address the reversibility. This focus on the method named simplified payment verification verifies the recorded transaction, which is signed by the

Elliptic Curve Digital Signature algorithm, which provides the private key to the sender who can check whether the fund is spent to the correct owner. This method also suggests the creation of wallet for the purpose of storing the essential keys, and another level of protection is made just by adding different devices in the security key for strengthening the security level of the transaction made by the owners [71]. The review of a blockchain-based systematic method is discussed in multiple domains. The investigation of status in blockchain technology is highlighted, and the characteristics needed for different applications are considered as a revolutionized technology for the usual business practices. The enabled application of blockchain is based on the systematic reviews and thematic content analysis. The blockchain technology can be classified for different applications such as data management, key themes, trends, and emerging areas for research. The shortcomings in these particular areas are identified and how this limitation varies between industries is discussed. The research gaps found in exploratory directions are also given for the researchers and academicians [72].

13.4   CONCLUSION

The blockchain through 5G technology, in combination with the IoT and IIoT, is discussed in this section. Besides, the innumerable applications of blockchain in the field of modern industrial sciences are also discussed. At the end, the administration of 5G along with the reconciliation of blockchain using IoT gadgets turns into a theme-changing scheme. It gives bits of knowledge about the mechanical uses of blockchain in 5G-empowered IoT gadgets. Here, the gratified data are partitioned, such as blockchain, IoT, IIoT, and 5G, quickly followed by separate modern applications. Mostly open issues and research challenges have been secured primarily for modern industrial applications. Inferable from the top of the line equipment prerequisites and the absence of similarity for large system network, utilization of the advances shrouded right now a typical stage is still a long way from reality. The more significant part of the mechanical applications has been canvassed; blockchain will be utilized to keep up with the rapid security information stream. Regardless of the past little scope of advancements in organizations with explicit applications, a lot of innovative research is required to address the particular requests relating to the coordinated effort of these advances. Among these lines, a relative examination of the current blockchain- with mechanical applications is performed with respect to the premise of explicit parameters.

REFERENCES

1. Alladi, Tejasvi, et al. "Block chain applications for Industry 4.0 and industrial IoT: A review." *IEEE Access* 7 (2019): 176935–176951.
2. Zhou, Keliang, Taigang Liu, and Lifeng Zhou, "Industry 4.0: Towards future industrial opportunities and challenges," in *2015 12th International Conference on Fuzzy Systems and Knowledge Discovery (FSKD)*. IEEE, 2015, pp. 2147–2152.
3. Davies, Ron. "Industry 4.0 digitalisation for productivity and growth." *European Parliamentary Research Service* 10 (2015).

4. Shrouf, Fadi, Joaquin Ordieres, and Giovanni Miragliotta., "Smart factories in Industry 4.0: A review of the concept and of energy management approached in production based on the Internet of Things paradigm," in *2014 IEEE International Conference on Industrial Engineering and Engineering Management*. IEEE, 2014, pp. 697–701.

5. Wu, Jinsong, et al. "Information and communications technologies for sustainable development goals: State-of-the-art, needs and perspectives." *IEEE Communications Surveys & Tutorials* 20.3(2018): 2389–2406.

6. Weyer, Stephan, Mathias Schmitt, Moritz Ohmer, and Dominic Gorecky, "Towards Industry 4.0-standardization as the crucial challenge for highly modular, multivendor production systems." *IFAC-Papersonline* 48.3(2015): 579–584.

7. M. Weyrich and C. Ebert, "Reference architectures for the Internet of Things." *IEEE Software* 33.1(2015): 112–116.

8. Salman, Ola, et al. "An architecture for the Internet of Things with decentralized data and centralized control." *2015 IEEE/ACS 12th International Conference of Computer Systems and Applications (AICCSA)*. IEEE, 2015. pp. 1–8.

9. Ansari, Rafay Iqbal, et al. "A new dimension to spectrum management in IoT empowered 5G networks." *IEEE Network* 33.4(2019): 186–193.

10. Almadhoun, Randa, et al. "A user authentication scheme of IoT devices using blockchain-enabled fog nodes." *2018 IEEE/ACS 15th International Conference on Computer Systems and Applications (AICCSA)*. IEEE, 2018.

11. Mistry, Ishan, et al. "Blockchain for 5G-enabled IoT for industrial automation: A systematic review, solutions, and challenges." *Mechanical Systems and Signal Processing* 135 (2020): 106382.

12. Miraz, Mahdi H. and Maaruf Ali. "Blockchain enabled enhanced IoT ecosystem security," in: *International Conference for Emerging Technologies in Computing*. Springer, 2018, pp. 38–46.

13. Hussain Shah, Sajjad and Ilyas Yaqoob. "A survey: Internet of Things (IOT) technologies, applications and challenges," in *2016 IEEE Smart Energy Grid Engineering (SEGE)*. IEEE 2016, pp. 381–385.

14. Yadav, Er Pooja, Er Ankur Mittal, and Hemant Yadav. "IoT: Challenges and issues in Indian perspective," in: *2018 3rd International Conference on Internet of Things: Smart Innovation and Usages (IoT-SIU)*, IEEE, 2018, pp. 1–5.

15. Pundir, Yogita, Nancy Sharma, and Yaduvir Singh. "Internet of things (IoT): Challenges and future directions." *International Journal of Advanced Research in Computer and Communication Engineering* 5.3(2016): 960–964.

16. Waleed, Ejaz, et al., "Internet of Things (IoT) in 5G wireless communications." *IEEE Access* 4 (2016): 10310–10314.

17. Dewey, Josias N., Robert Hill, and Rebecca Plasencia, "Blockchain and 5G-enabled Internet of Things (IOT) will redefine supply chains and trade finance." *Secured Lender* (2018): pp. 42–45.

18. Skouby, Knud Erik and Per Lynggaard, "Smart home and smart city solutions enabled by 5G, IoT, AAI and CoT services," in: *2014 International Conference on Contemporary Computing and Informatics (IC3I)*. 2014, IEEE pp. 874–878.

19. Roshan, Rakesh and Abhay Kr Ray. "Challenges and risk to implement IOT in smart homes: An Indian perspective." *International Journal of Computer* 153 (2016): 16–19.

20. Lazaroiu, Cristian and Mariacristina Roscia, "Smart district through IoT and blockchain," in: *2017 IEEE 6th International Conference on Renewable Energy Research and Applications (ICRERA)*. 2017, pp. 454–461.

21. Zanella, Andrea, et al. "Internet of Things for smart cities." *IEEE Internet of Things* 1.1(2014): 22–32.

22. Talari, Saber, et al. "A review of smart cities based on the Internet of Things concept." *Energies* 10.4(2017): 421.

23. Cenedese, Angelo, et al. "Padova smart city: An urban Internet of Things experimentation," in: *Proceeding of IEEE International Symposium on a World of Wireless, Mobile and Multimedia Networks*. 2014, IEEE pp. 1–6.
24. Pham, Thanh Nam, et al. "A cloud-based smart-parking system based on Internet-of-Things technologies." *IEEE Access* 3 (2015): pp. 1581–1591.
25. Kumari, Aparna, et al. "Fog computing for Healthcare 4.0 environment: Opportunities and challenges." *Computers & Electrical Engineering* 72 (2018): 1–13.
26. Gapchup, Akshay, et al. "Health care systems using Internet of Things." *IJIRCCE* 4 (2016): 12.
27. Islam, SM. Riazul, "The Internet of Things for health care: A comprehensive survey." *IEEE Access* 3 (2015): 678–708.
28. Baker, Stephanie B., Wei Xiang, and Ian Atkinson. "Internet of Things for smart healthcare: Technologies, challenges, and opportunities." *IEEE Access* 5 (2017): 26521–26544.
29. Hathaliya, Jigna J., et al. "Securing electronics healthcare records in Healthcare 4.0: A biometric-based approach." *Computers & Electrical Engineering* 76 (2019): 398–410.
30. What is an electronic health record (EHR)? URL:https://www.healthit.gov/faq/what-electronic-health-record-ehr.
31. Ekblaw, Ariel, et al. "A case study for blockchain in healthcare: "MedRec" prototype for electronic health records and medical research data." *Proceedings of IEEE Open & Big Data Conference*. Vol. 13, 2016. IEEE, p. 13.
32. Viriyasitavat, Wattana, et al. "Blockchain-based business process management (BPM) framework for service composition in Industry 4.0." *Journal of Intelligent Manufacturing* (2018): 1–12.
33. Xu, Li Da, Eric L. Xu, and Ling Li. "Industry 4.0: State of the art and future trends." *International Journal of Production Research* 56.8(2018): 2941–2962.
34. Kapitonov, Aleksandr, et al. "Blockchain based protocol for economical communication in industry 4.0." *2018 Crypto valley Conference on Blockchain Technology (CVCBT)*. IEEE, 2018. pp. 41–44.
35. Kapitonov, Aleksandr, et al. "Blockchain-based protocol of autonomous business activity for multi-agent systems consisting of UAVs." *2017 Workshop on Research, Education and Development of Unmanned Aerial Systems (RED-UAS)*. IEEE, 2017. pp. 84–89.
36. AIRA – open source software for smart cities and Industry 4.0 projects. URL:https://aira.life (visited on 02/12/2019).
37. UNCHAINET – decentralized cloud platform. URL:https://www.unchainet.com/ (visited on 02/17/2019).
38. What is supply chain (SC)? URL:https://whatis.techtarget.com/definition/supply-chain (visited on 02/17/2019).
39. Lin, Jun, et al. "Blockchain and IoT based food traceability for smart agriculture." *Proceedings of the 3rd International Conference on Crowd Science and Engineering*. 2018. pp. 1–6.
40. Tripoli, Mischa, and Josef Schmidhuber. "Emerging opportunities for the application of blockchain in the *Agri*-food industry." *FAO and ICTSD: Rome and Geneva*. Licence: CC BY-NC-SA3, 2018. pp. 1–40.
41. Yuan, Yong and Fei-Yue Wang. "Towards blockchain-based intelligent transportation systems." *2016 IEEE 19th International Conference on Intelligent Transportation Systems (ITSC)*. IEEE, 2016. pp. 2663–2668.
42. Applications of Unmanned Aerial Vehicle (UAV) based remote sensing in NE region – ISRO. URL: https://www.isro.gov.in/applications-of-unmannedaerial-vehicle-uav-based-remote-sensing-ne-region.

43. Chandrasekharan, Sathyanarayanan, et al. "Designing and implementing future aerial communication networks." *IEEE Communications Magazine* 54.5(2016): 26–34.

44. Lin, Xingqin, et al. "The sky is not the limit: LTE for unmanned aerial vehicles." *IEEE Communications Magazine* 56.4(2018): 204–210.

45. Sedjelmaci, Hichem, Sidi Mohammed Senouci, and Mohamed-Ayoub Messous. "How to detect cyber-attacks in unmanned aerial vehicles network?" *2016 IEEE Global Communications Conference (GLOBECOM)*. IEEE, 2016. pp. 1–6.

46. Banerjee, Mandrita, Junghee Lee, and Kim-Kwang Raymond Choo. "A blockchain future for Internet of Things security: A position paper." *Digital Communications and Networks* 4.3(2018): 149–160.

47. Kishigami, Junichi, et al. "The blockchain-based digital content distribution system." *2015 IEEE Fifth International Conference on Big Data and Cloud Computing*. IEEE, 2015. pp. 187–190.

48. Xu, Ruzhi, et al. "Design of network media's digital rights management scheme based on blockchain technology." *2017 IEEE 13th International Symposium on Autonomous Decentralized System (ISADS)*. IEEE, 2017. pp. 128–133.

49. Bhowmik, Deepayan and Tian Feng. "The multimedia blockchain: A distributed and tamper-proof media transaction framework." *2017 22nd International Conference on Digital Signal Processing (DSP)*. IEEE, 2017. pp. 1–5.

50. Khezr, Seyednima, et al. "Blockchain technology in healthcare: A comprehensive review and directions for future research." *Applied Sciences* 9.9(2019):Pp: 1736.

51. Rennock, Michael, Alan Cohn, and Jared R. Butcher. "Blockchain technology and regulatory investigations." Practical Law Litigation (2018). pp. 35–44.

52. Dorri, Ali, Salil S. Kanhere, and Raja Jurdak. "Blockchain in Internet of Things: Challenges and solutions." *arXiv preprint arXiv:*1608.05187 (2016).

53. Yaeger, Kurt, et al. "Emerging blockchain technology solutions for modern healthcare infrastructure." *Journal of Scientific Innovation in Medicine* 2.1(2019).

54. Pilkington, Marc and Jong-Hyouk Lee. "How the blockchain revolution will reshape the consumer electronics industry." *forthcoming (provisional draft, do not cite)* (2017).

55. Bernabe, Jorge Bernal, et al. "Privacy-preserving solutions for blockchain: Review and challenges." *IEEE Access* 7 (2019): 164908–164940.

56. Gaggioli, Andrea. "Blockchain technology: Living in a decentralized everything." *Cyberpsychology, Behavior, and Social Networking* 21.1(2018): 65–66.

57. Leible, Stephan, et al. "A review on blockchain technology and blockchain projects fostering open science." *Frontiers in Blockchain* 2 (2019): 16.

58. Mahmoud, Qusay H., Michael Lescisin, and May AlTaei. "Research challenges and opportunities in blockchain and cryptocurrencies." *Internet Technology Letters* 2.2(2019): e93.

59. Al-Fuqaha, Ala, et al. "Internet of Things: A survey on enabling technologies, protocols, and applications." *IEEE Communications Surveys & Tutorials* 17.4(2015): 2347–2376.

60. MIT Technology review: The blockchain issue. (2018) Available: https://www.technologyreview.com/magazine/2018/05/.

61. Subramanian, Hemang. "Decentralized blockchain-based electronic marketplaces." *Communications of the ACM* 61.1(2017): 78–84.

62. Peters, Gareth W. and Efstathios Panayi. "Understanding modern banking ledgers through blockchain technologies: Future of transaction processing and smart contracts on the Internet of money." *Banking Beyond Banks and Money*. Springer, Cham, 2016. 239–278.

63. Tapscott, Alex and Don Tapscott. "How blockchain is changing finance." *Harvard Business Review* 1.9(2017): 2–5.

64. Christidis, Konstantinos and Michael Devetsikiotis. "Blockchains and smart contracts for the Internet of Things." *IEEE Access* 4 (2016): 2292–2303.

65. De Filippi, Primavera. "The interplay between decentralization and privacy: The case of blockchain technologies." *Journal of Peer Production* 7(2016). p. 7.
66. Conoscenti, Marco, Antonio Vetro, and Juan Carlos De. "Blockchain for the Internet of Things: A systematic literature review," in: *Proceedings of the 2016 IEEE/ACS 13th International Conference of Computer Systems and Applications (AICCSA)*. 2016, IEEE pp. 1–6.
67. Fernández-Caramés, Tiago M. and Paula Fraga-Lamas. "A review on the use of blockchain for the Internet of Things." *IEEE Access* 6 (2018): 32979–33001.
68. Puthal, Deepak, et al. "Everything you wanted to know about the blockchain: Its promise, components, processes, and problems." *IEEE Consumer Electronics Magazine* 7.4(2018): 6–14.
69. Madumidha, S., Ranjani, P.S., Vandhana, U. and Venmuhilan, B. "A theoretical implementation: Agriculture-food supply chain management using blockchain technology." *2019 TEQIP III Sponsored International Conference on Microwave Integrated Circuits, Photonics and Wireless Networks (IMICPW)*. IEEE, 2019. pp. 174–178.
70. Alamri, Malak, N.Z. Jhanjhi, and Mamoona Humayun. "Blockchain for Internet of Things (IoT) research issues challenges & future directions: A review." *International Journal of Computer Science and Network Security* 19 (2019): 244–258.
71. bt Abd Halim, Norul Suhaliana, et al. "Blockchain security hole: Issues and solutions," *International Conference of Reliable Information and Communication Technology*. Springer, Cham, 2017. pp. 739–746.
72. Casino, Fran, Thomas K. Dasaklis, and Constantinos Patsakis. "A systematic literature review of blockchain-based applications: Current status, classification and open issues." *Telematics and Informatics* 36 (2019): 55–81.

Index